Advanced Sciences and Technologies for Security Applications

Indexed by SCOPUS

The series Advanced Sciences and Technologies for Security Applications comprises interdisciplinary research covering the theory, foundations and domain-specific topics pertaining to security. Publications within the series are peer-reviewed monographs and edited works in the areas of:

- biological and chemical threat recognition and detection (e.g., biosensors, aerosols, forensics)
- crisis and disaster management
- terrorism
- cyber security and secure information systems (e.g., encryption, optical and photonic systems)
- traditional and non-traditional security
- energy, food and resource security
- economic security and securitization (including associated infrastructures)
- transnational crime
- human security and health security
- social, political and psychological aspects of security
- recognition and identification (e.g., optical imaging, biometrics, authentication and verification)
- smart surveillance systems
- applications of theoretical frameworks and methodologies (e.g., grounded theory, complexity, network sciences, modelling and simulation)

Together, the high-quality contributions to this series provide a cross-disciplinary overview of forefront research endeavours aiming to make the world a safer place.

The editors encourage prospective authors to correspond with them in advance of submitting a manuscript. Submission of manuscripts should be made to the Editor-in-Chief or one of the Editors.

Reza Montasari · Victoria Carpenter ·
Anthony J. Masys
Editors

Digital Transformation in Policing: The Promise, Perils and Solutions

 Springer

Editors
Reza Montasari
Department of Criminology, Sociology
and Social Policy, School of Social
Sciences
Swansea University
Swansea, Wales, UK

Victoria Carpenter
University of Bedfordshire
Luton, UK

Anthony J. Masys
College of Public Health
University of South Florida
Tampa, FL, USA

ISSN 1613-5113   ISSN 2363-9466 (electronic)
Advanced Sciences and Technologies for Security Applications
ISBN 978-3-031-09693-8   ISBN 978-3-031-09691-4 (eBook)
https://doi.org/10.1007/978-3-031-09691-4

This Springer imprint is published by the registered company Springer Nature Switzerland AG
The registered company address is: Gewerbestrasse 11, 6330 Cham, Switzerland

Contents

Digitizing Policing: From Disruption to Innovation Through Futures Thinking and Anticipatory Innovation

Anthony J. Masys

Abstract In today's security landscape we find ourselves confronted with problems of dynamic complexity, uncertainty and ambiguity with disruptive technologies being exploited in the criminal domain. Such key technological developments such as Artificial Intelligence (AI), machine learning algorithms, quantum computing, 5G, Dark web networks and cryptocurrencies, the Internet of All Things, 3D printing, molecular biology and genetics are all influencing the criminal landscape and shaping the future digitalization of law enforcement capabilities. For policing, these strategic challenges require a strategic response that is rooted in innovation and agility. As described in Policing in the UK, '…policing will need to get better at anticipating emerging threats, think more innovatively about the best policies and interventions for addressing them and mobilise responses quickly to maximise chances of success (College of Policing in Policing in England and Wales future operating environment 2040. https://paas-s3-broker-prod-lon-6453d964-1d1a-432a-9260-5e0ba7d2fc51.s3.eu-west-2.amazonaws.com/s3fs-public/2020-08/Future-Operating-Environment-2040_0.pdf, 2020, p. 3). Moving towards the development of solutions to this complex problem space characterized by disruptive technologies for criminality depends on the lens we use to examine them and how we frame the problem. Futures thinking and foresight analysis figure prominently in shaping law enforcements ability to understand how emerging technologies and criminality intersect and thereby how law enforcement can position itself for the future through anticipatory innovation. This chapter introduces futures thinking and foresight as problem framing approaches that will inform the digitalization of the law enforcement domain.

Keywords Problem framing · Systems thinking · Anticipatory innovation · Futures thinking

A. J. Masys (✉)
International Centre for Policing and Security, USW, Pontypridd, UK
e-mail: Anthony.Masys@gmail.com

1 Introduction

In today's security landscape we find ourselves confronted with problems of dynamic complexity, uncertainty and ambiguity with disruptive technologies being exploited in the criminal domain. Such key technological developments such as Artificial Intelligence (AI), machine learning algorithms, quantum computing, 5G, Dark web networks and cryptocurrencies, the Internet of All Things, 3D printing, molecular biology and genetics are all influencing the criminal landscape and shaping the future digitalization of law enforcement capabilities.

The societal influence technology brings not only includes benefits but also increased risk exposure and greater threats. This society/technology nexus has become a focal point for cybercrime, cyber attacks and cyber warfare. Moving towards the development of solutions to this complex problem space characterized by disruptive technologies for criminality depends on the lens we use to examine them and how we frame the problem. Futures thinking and foresight analysis figure prominently in shaping law enforcement's ability to understand how emerging technologies and criminality intersect and thereby how law enforcement can position itself for the future through the application of anticipatory innovation.

2 Disruption

The societal safety and security landscape is being challenged by the emergence of disruptive technologies supporting all forms of criminality. This transformation of the criminal and terrorist landscape with the exploitation of drones and the dark web has already challenged law enforcement. As described in [14], 'Criminals are no longer restricted to burner phones, guns, and getaway cars. They're finding new, sophisticated ways to smuggle contraband, conduct counter-surveillance, and retaliate against those who threaten their schemes'. Drones, for example have disrupted the criminal landscape by providing advanced capabilities to support surveillance, smuggling and unmanned weapon systems. Similarly, drones have been employed to advance terrorist agendas. As described by Liang [14],

> Since 2016, ISIS has been using drones to carry out intelligence, surveillance, and reconnaissance missions. ISIS also conducted attacks with drones carrying explosives which were used to kill enemy combatants in northern Iraq. ISIS formed an "Unmanned Aircraft of the Mujahedeen" unit. In 2017 ISIS boasted that its drone attacks had killed or wounded 39 soldiers in one week. In 2017, ISIS pinned down Iraqi security forces during one 24-hour period in Syria using commercial and homemade drones executing 70 drone missions. ISIS has also distributed online guidance on the use of drones as well as propaganda calling for attacks involving drones. In September 2017, FBI Director Christopher Wray told the Senate that drones constituted an imminent terrorist threat to American cities.

As a disruptive technology, drones pose novel and difficult problems for law enforcement and security establishments. Similarly, developments along the areas of Artificial Intelligence (AI), quantum computing, 5G, Dark web networks and cryptocurrencies, the Internet of All Things, 3D printing, molecular biology and genetics will shock the law enforcement domain and require anticipatory foresight and design thinking to manage the law enforcement innovation required. A detailed description of such disruptive technologies impact and intersection with criminality can be found in [9]. Disruption occurs at the intersection of new emerging technologies and the unseen applications in the criminal landscape thereby opening the door to new forms of criminality. For example,

> In recent years, technology has acted as a driver of innovation across the whole spectrum of criminality, with criminals quickly adopting new modi operandi and activities enabled by advanced technologies. The emergence of the online trade in illicit goods and services has created entirely new criminal markets. Virtual currencies and alternative banking platforms are enabling the rapid flow of criminal finances, and new communication technologies, including the use of encrypted communications, have enabled criminals and terrorists to connect and interact covertly [9, p. 9].

The novel forms of criminality that are emerging are exploiting new societal dynamics and developments such as the Smart City and our wireless society. This nexus of society and technology is not without its vulnerabilities. For example, the COVID-19 pandemic revealed global and national societal vulnerabilities and systemic risks [30] that were exploited resulting in an increase in Cyber threats of 81% since the global pandemic [11]. WEF Global Risks Report [31] ranks cyber risks as among the top global risks. In one estimate, the global cost of cyber crime by 2025 is said to be upwards of $10 Trillion.[1] With the growth in the applications of cyberspace '…new market places for the sale and exchange of illegal weapons and drugs, other illicit materials and even the trafficking and exploitation of human beings…' [4, p. v] has emerged. Recent cybersecurity events have highlighted and elevated the profile of cyber risk and systemic cyber risk described in [18]. This points to the requirement for new ways of thinking regarding threat and risk analysis. Relying on the linear extrapolation of historical data will not suffice. What is required is greater foresight linked with innovation.

As described by Reez [23], traditional mindsets and practices are inadequate to deal with such disruption. The application of futures thinking and foresight opens a space to work under conditions of volatility, uncertainty, complexity and ambiguity (VUCA). It allows us to reframe, to view the possible and plausible and create conditions to support anticipatory innovation within a learning culture. Foresight helps to distinguish signals from noise and to better prepare for uncertainty and disruption. Futures thinking and foresight methodologies to support anticipatory innovation in the policing and security landscape thereby emerges as part of the solution space.

---

[1] https://securityboulevard.com/2021/03/cybercrime-to-cost-over-10-trillion-by-2025/.

3   Futures Thinking

The application of futures thinking is a strategic initiative that can have influence across the tactical and operational domains. With applied futures thinking, the future becomes a creative landscape in which to influence and shape the future through anticipatory innovation. Futures thinking is about exploring uncertainty, considering multiple futures, exploring intended and unintended consequences, seeking out multiple perspectives, and exploring assumptions. Futures thinking is not about predicting the future but recognizes that there are a range of possible futures. These futures can be shaped and influenced by the decisions and actions we take today. Hence combining futures thinking with systems thinking and design thinking operationalizes the process of anticipatory innovation (Fig. 1).

Anticipatory innovation is about helping to shape how the future might play out, rather than being forced to respond to it when it arrives. The methodologies used to support anticipatory innovation are designed for VUCA conditions under which the future unfolds. With that in mind, assumption based planning [7, 15] becomes a critical tool in the futures thinking tool box to support social learning for anticipatory innovation. Through assumptions based planning, one creates scenarios and explicitly articulates the assumptions that bound the scenario. With any scenario, some assumptions are correct, while some are volatile and may become invalid. With this in mind, assumption based planning facilitates a mindset of reflective practices pertaining to scenario assumptions thereby identifying those assumptions that are load bearing. In so doing, futures thinking challenges and exposes implicit assumptions and dominant perspectives, to explore surprises and disruptions. Futures thinking is apropos for both anticipating future change, and for responding and shaping future change. As noted in [32] 'Thinking through multiple future scenarios

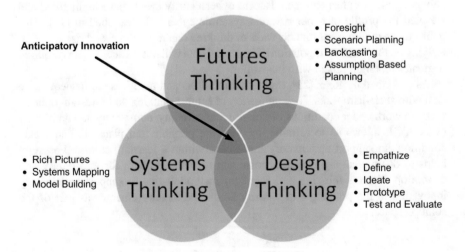

Fig. 1 Anticipatory innovation

today, allows us to prepare, both mentally and institutionally, for emerging and alternative futures tomorrow'.

Futures thinking through foresight '…uses a range of methodologies, such as scanning the horizon for emerging changes, analysing weak signals and megatrends, and developing multiple scenarios, to reveal and discuss useful ideas about the future' [20]. For example as detailed in [20]:

Horizon scanning: seeking and researching signals of change in the present and their potential future impacts. Horizon scanning is the foundation of any strategic foresight process. It can involve desk research, expert surveys, and review of existing futures literature.

Megatrends analysis: exploring and reviewing of large-scale changes building in the present at the intersection of multiple policy domains, with complex and multidimensional impacts in the future.

Scenario planning: developing multiple stories or images of how the future could look in order to explore and learn from them in terms of implications for the present.

Visioning and back-casting: developing an image of an ideal (or undesirable) future state, and working backwards to identify what steps to take (or avoid).

Futures thinking thereby embraces '…multi-method approaches allowing a plurality of voices to frame problems and iteratively unfold plausible assessments' [23, p. 332]. It follows that through futures thinking through the application of design thinking, the law enforcement community can better position itself to apply anticipatory innovation for the digitalization of policing thereby combating security threats proactively and engage in prevention strategic interventions [9, p. 3].

4 Discussion

The four characteristics of VUCA are shaping the policing and law enforcement landscape in terms of not only events but also with regards to our understanding and sensemaking of security events and implications [17]. As described in [13], the components are defined as:

Volatility - Volatility refers to the speed of change in an industry, market or the world in general. It is associated with fluctuations in demand, turbulence and short time to markets and it is well-documented in the literature on industry dynamism. The more volatile the world is, the more and faster things change.

Uncertainty - Uncertainty refers to the extent to which we can confidently predict the future. Part of uncertainty is perceived and associated with people's inability to understand what is going on. Uncertainty, though, is also a more objective characteristic of an environment. Truly uncertain environments are those that don't allow any prediction, also not on a statistical basis. The more uncertain the world is, the harder it is to predict.

Complexity - Complexity refers to the number of factors that we need to take into account, their variety and the relationships between them. The more factors, the greater their variety and the more they are interconnected, the more complex an environment is. Under high complexity, it is impossible to fully analyze the environment and come to rational conclusions. The more complex the world is, the harder it is to analyze.

Ambiguity - Ambiguity refers to a lack of clarity about how to interpret something. A situation is ambiguous, for example, when information is incomplete, contradicting or too inaccurate to draw clear conclusions. More generally it refers to fuzziness and vagueness in ideas and terminology. The more ambiguous the world is, the harder it is to interpret.

Within this VUCA world, societal vulnerabilities are often hardwired into our national and global interdependencies and lie dormant until triggered by some shock [30]. Situational contexts combined with technological innovations often reveal societal vulnerabilities that can be exploited for nefarious gains. For example, across the landscape of the future where we see greater societal/technology integration with the cyber interdependencies and advent of AI, so do the risks of criminal exploitation. As described in [5]:

> Opportunities for AI-enabled crime exist both in the specifically computational domain (overlapping with traditional notions of cybersecurity) and also in the wider world. Some of these threats arise as an extension of existing criminal activities, while others may be novel. To adequately prepare for and defend against possible AI threats, it is necessary to identify what those threats might be, and in what ways they are likely to impact our lives and societies.

Further on the AI front, Abaimov and Martellini [1, p. 8] argue that, 'The cyber world has been both enhanced and endangered by AI. On one side, the performance of the existing security services has been improved, and new tools created. On the other hand, it generates new cyber threats both through enhancing the attacking tools and through its own imperfection and vulnerabilities. The new AI-backed attacks and malware like DeepHack (framework), DeepLocker (ransomware) or Deep Exploit can be named as examples'. With these new and enhanced capabilities, the impact of attackers has reached new levels of significance. From a criminal exploitation perspective, Abaimov and Martellini [1, pp. 115–116] argue that:

> AI greatly improves the attackers' capabilities in passive and active reconnaissance, generation of exploitation payloads, traffic masquerading, creating phishing emails, or causing physical damage to systems. Cyber attacks are of particular danger to autonomous systems when their operational process may be manipulated. The number of AI-powered attack tools will only grow in the future, as machine learning is gaining popularity and accessibility.

This supports the requirement that law enforcement must go beyond lessons learned from incidents to also learning from the future. Futures thinking combined with systems thinking and design thinking supports this anticipatory innovation posture and creates a learning opportunity from which new insights can be garnered to shape the future.

5 Anticipatory Innovation

The challenges for the policing landscape are manifest. With a rapidly changing and dynamic societal environment coupled with the volatility and complex problems associated with crime, the evolving problem space looks toward the requirement for methodologies that can shape the future rather than just reacting to it: to be able to anticipate different (probable, plausible and possible) futures and prepare for these realities. This is the domain of anticipatory innovation.

As described in [29, 31] 'Anticipation does not mean predicting the future, but rather it is about asking questions about plausible futures so that we may act in the present to help bring about the kind of futures we decide we want...It is a capacity connected to engaging with alternative futures, based on sensitivity to weak signals, and an ability to visualize their consequences, in the form of multiple possible outcomes ...'.

6 Anticipatory Innovation Mindset and Methodologies

To enable Anticipatory innovation calls for new mindsets, tools and approaches that taps into the creativity and imagination necessary to navigate the VUCA conditions that characterize the criminal landscape. **Anticipatory innovation lies at the nexus of futures thinking, design thinking and systems thinking** (Fig. 1).

Scenario planning/analysis [15] emerge as one of the key toolsets amongst the futures thinking approaches that explores the cone of plausibility (Fig. 2). Such an approach creates a shared space for participatory engagement and learning that leverages a variety of perspectives and voices.

7 Futures Thinking

Futures thinking creates a lens into the complexity of the future. Exploring the plausibility space requires us to step into a space of strategic and creative thinking that leverages reflective practices, learning and knowledge creation. As described in [23, p. 330] 'Prospective leaders have to get familiar with what is called

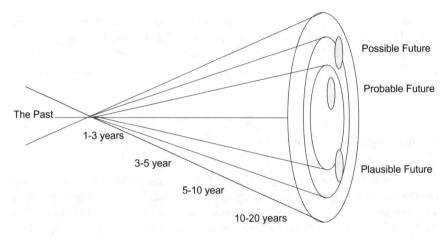

Fig. 2 Cone of plausibility

visioning, scenario building [34], weak signals [24] wild cards [27], hidden influences, horizon scanning, action learning storytelling [22]. Embracing a mindset of exploratory futuring creates an opportunity for social learning. To operationalize futures thinking, Reez [23, p. 335] argues that 'Effective foresight requires open mindedness, broad thinking, stakeholder dialogue, multiple communicative loops and abductive reasoning. …Future oriented analysis therefore needs to emphasize processes that support insight, intuition and innovation, instead of relying on historical data'.

Futures thinking and foresight not only use historical data and analytics but leverages creativity, imagination and experimentation from across a wide range of participants. As described in [3, p. 3] '…it does not look only at what is possible but at what is desired. In this way, futures thinking and foresight are different from traditional forecasting, which is narrowly focused. Because they are participatory, futures thinking and foresight strengthen cross-sectoral links, encourage the emergence of integrated solutions, and empower people to create the future they desire.

For a detailed explanation of the cone see [32].

The application of futures thinking can be applied to 2 exploratory domains followed by an integration of these perspectives to map out the influences from the interdependencies and interconnectivity.

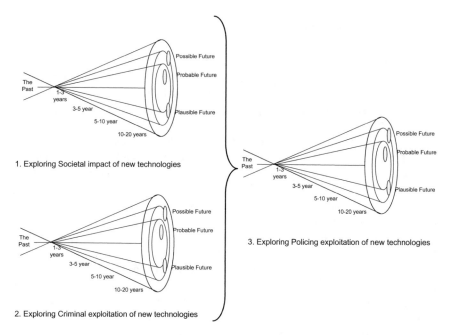

1. Exploring Societal impact of new technologies

3. Exploring Policing exploitation of new technologies

2. Exploring Criminal exploitation of new technologies

The analysis begins with an exploration of the societal impacts of new technologies in order to reveal vulnerabilities. Step 2 is an exploration of the exploitation of new technologies by criminal organizations. Combining these 2 steps allows the researcher to contextually situate the technologies showing the intersection of criminality, technology and societal vulnerability and fragility. Step 3 explores this landscape through the policing context, identifying strategic intervention strategies that can be implemented in the present to influence the future.

Scenario thinking plays a key role in this analysis with a focus on thinking the unthinkable. As described in [15],

> Black swans represent the unpredictable. They represent "[…] our misunderstanding of the likelihood of surprises" [27, p. 2]. A black swan is described by Taleb [27] as that which is an outlier, that which is outside the realm of regular expectations which carries with it an extreme impact such as natural disasters, market crashes, catastrophic failure of complex socio-technical systems and terrorist events such as 9/11. These "surprising events" reflect an organizations inability to recognize evidence of new vulnerabilities or the existence of ineffective countermeasures [33, p. 24]. This necessitates the requirement to readjust to their existence and thereby the need to consider the extremes [27, p. xx].
>
> With consideration of emerging and systemic risks and inherent uncertainty associated with surprising events, planning for and managing risk, crisis and disasters requires understanding of the space of possibilities in order to avoid unrealistic expectations that can influence the management of disasters and catastrophes…. A scenario is essentially a story, describing potential future conditions and their emergence to facilitate sense making and to inform decision making. The thought process involved in scenario planning supports "thinking the unthinkable" exploring uncertainty and challenging mental models and assumptions in order to recognize alternate futures in a space of possibilities'.

8 Systems Thinking

Futures thinking looks to explore and understand the driving factors and influences shaping possible scenarios. Supporting this is the approach of systems thinking. Angel Gurria, the OECD Secretary General declared in March 2019 that, 'unless we adopt a systems thinking approach, unless we employ systems thinking, we will fail to understand the world we are living in' [12, p. 641].

As described in Masys [19] systems thinking emerges as both a worldview and a process in the sense that it informs ones understanding regarding a system and can be used as an approach in problem solving [8, p. 5]. 'Systems thinking' as discussed in [25] emphasizes interconnectedness, causal complexity and the relation of parts to the whole [2], thereby challenging traditional linear and reductionist thinking and simple causal explanations. Feedback and feedforward loops emerge from the analysis thereby giving insights into intended and unintended consequences from decision and actions that influence system behavior in response to exogenous and endogenous shocks. This is well demonstrated and explained within the context of the global impact of COVID-19 and system risks [30]. Systems thinking purports that, although events and objects may appear distinct and separate in space and time, they are all interconnected. As a process, systems thinking recognizes the requirement to assess the system within its environment and context [26].

Given the conditions of volatility, uncertainty, complexity and ambiguity that characterizes the policing landscape, quick-fix solutions are ill-suited. As described in [12, p. xx), 'in its most advanced form, systems approaches encourages the employment of a variety of methodologies in combination to manage 'messes' and 'wicked problems'. Critical systems practice informs this way of working and demonstrates how decision makers can achieve successful outcomes by becoming 'multimethodological'. Hence, systems thinking is clearly an underpinning foundation supporting the application of foresight. As described in [10, p. 145], 'applying a systemic lens to complex problems can help map the dynamics of the system, explore the ways in which the relationships between system components affect its functioning, and ascertain which interventions can lead to better results.

With strategic interventions and initiatives being deployed by law enforcement agencies, systems thinking helps to support our understanding of the issues, reflect and test the consequences of decisions and intervention strategies, thereby identifying and understanding critical relationships in the system and to reduce unintended consequences of interventions and create better agility.

9 Design Thinking

Design Thinking coupled with futures thinking and systems thinking is solution oriented methodology used to solve complex problems [16]. In its most basic form it involves a combination of imagination, analysis and creativity to bring about solution focused and action oriented results. To facilitate this, the design thinking approach is rooted in a learning environment.

Through the phases of Inspiration, Ideation and Implementation, Design Thinking is operationalized through an iterative (not linear) 5 step process (Fig. 3). This process is described in detail (http://dschool.stanford.edu/redesigningtheater/the-design-thinking-process/).

EMPATHIZE: Work to fully understand the experience of disruptive technologies in across the societal, criminal and law enforcement landscape. Do this through observation, interaction, and immersing yourself in their experiences.

DEFINE: Process and synthesize the findings from the empathy work in order to form a point of view that you will address with your design. With this comes the understanding of the interdependencies that characterize the landscape.

IDEATE: Explore a wide variety of possible solutions through generating a large quantity of diverse possible solutions, allowing you to step beyond the obvious and explore a range of ideas.

PROTOTYPE: Transform your ideas into an operational form so that you can experience and interact with them and, in the process, learn and develop more empathy.

TEST: Try out high-resolution concepts and use observations and feedback to refine prototypes, learn more about the criminal problem space, and refine your original point of view.

The application of design thinking supporting anticipatory innovation is critical. It brings a solution and action oriented lens to the problem space. As described in [21] '…designers have specific abilities to produce novel unexpected solutions, tolerate uncertainty, work with incomplete information, apply imagination and forethought to practical problems and use drawings and other modeling media as means to problem solving. He further argues that designers must be able to resolve ill-defined problems, adopt solution-focusing strategies, employ abductive/productive/appositional

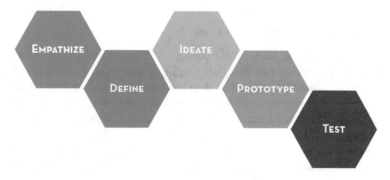

Fig. 3 Design thinking process

thinking and use non-verbal, graphic and spatial modeling media'. These qualities resonate with the VUCA conditions that describe the policing environment and the disruptive technologies that are emerging in future problem space.

Within this creative space of design thinking, it is critical to cultivate mindfulness, expanding our powers of imagination to conceive innovative solutions [23, p. 324]. With this in mind, there exist organizational cultural requirements to facilitate design thinking in support of anticipatory innovation.

The application of anticipatory innovation requires not only a variety of approaches and tools but also a certain strategic culture that fosters imagination, inquiry, analysis and creativity, thereby creating a 'safe space' for exploratory learning. Inherent storytelling is used to forge shared understanding and facilitate learning thereby challenging assumptions by shedding light on ambiguity, uncertainty, complexity. A culture that embraces diversity and inclusion is necessary to support sensemaking across various perspectives in this social learning journey.

Anticipatory Innovation (emerging at the intersection of futures thinking, systems thinking and design thinking) is a disruptive approach in and of itself. It entails the policing law enforcement agencies to:

- Embrace non-traditional data sets, diversity and inclusion of perspectives
- identify and test the assumptions
- explore beyond the assumed future
- generate new insights
- think through future intended and unintended consequences
- innovate and real-time.

With a cultural environment and the applied intersection of futures thinking, systems thinking and design thinking, the policing community is positioned to take action towards shaping desired futures. It is more than just creating new knowledge about plausible futures but is about innovating based on that information.

10   Conclusion

Institutionalizing the imaginative process

The futures thinking lens lends itself to anticipatory innovation to support strategic digitalization of policing to address the emerging technologies in the criminal landscape. The characteristics of futures thinking such as revealing and challenging mental models and assumptions, broad thinking, stakeholder dialogue, multiple communicative loops and abductive reasoning creates a reflective learning space. With systems thinking, we begin to explore the interdependencies and interconnectivity across systems to reveal the impact of decisions pertaining to intended and unintended consequences. This opens a space to what [28] calls **design-based** foresight as 'next generation foresight. …foresight should not try to predict or know the

future but create it. From that point of view, foresight- conceptualized as design-based is essentially action oriented. …action learning projects (cited in [23, p. 336].
As described in Europol [9, p. 6]:

Disruption through technological progress occurs as a result of the convergence of new emerging technologies, and the ways they challenge existing legal and regulatory frameworks with previously unseen applications. Disruption through new technologies presents both challenges and opportunities for law enforcement authorities through the emergence of new or significantly altered criminal activities as well as through the potential exploitation of these technologies by law enforcement authorities.

The intersection of futures thinking, systems thinking and design thinking facilitates the policing community with the tools and methodologies to explore, understand and prepare for the future in order to navigate, adapt, and shape the future through better policies [20, p. 3].

References

1. Abaimov S, Martellini M (2022) Machine learning for cyber agents: attack and defence. Springer Publishing
2. Ackoff R (1994) Systems thinking and thinking systems. Syst Dyn Rev 10(2–3):175–188
3. ADB (2020) APRIL 2020 futures thinking in Asia and the Pacific why foresight matters for policy makers
4. Akhgar B, Brewster B (eds) (2016) Combatting cybercrime and cyberterrorism: challenges, trends and priorities. Springer
5. Caldwell M, Andrews JTA, Tanay T, Griffin JD (2020) AI-enabled future crime. Crime Sci 9:14
6. College of Policing (2020) Policing in England and Wales future operating environment 2040. https://paas-s3-broker-prod-lon-6453d964-1d1a-432a-9260-5e0ba7d2fc51.s3.eu-west-2.amazonaws.com/s3fs-public/2020-08/Future-Operating-Environment-2040_0.pdf
7. Dewar JA (2002) Assumption based planning: a tool for reducing avoidable surprises. Cambridge University Press
8. Edson R (2008) Systems thinking. Applied: a primer. ASysT Institute. http://www.anser.org/docs/systems_thinking_applied.pdf
9. Europol (2019) Do criminals dream of electric sheep? How technology shapes the future of crime and law enforcement. https://www.europol.europa.eu/publications-events/publications/do-criminals-dream-of-electric-sheep-how-technology-shapes-future-of-crime-and-law-enforcement
10. Hynes W, Lees M, Müller J (eds) (2020) Systemic thinking for policy making: the potential of systems analysis for addressing global policy challenges in the 21st century. In: New approaches to economic challenges. OECD Publishing, Paris. https://doi.org/10.1787/879c4f7a-en
11. InfoSecurity (2019) https://www.infosecurity-magazine.com/news/81-orgs-cyber-threats-covid19/
12. Jackson M (2019) Critical systems thinking and the management of complexity. Wiley
13. Kraaijenbrink J (2018) What does VUCA really mean? Forbes. https://www.forbes.com/sites/jeroenkraaijenbrink/2018/12/19/what-does-vuca-really-mean/?sh=6fdda5c917d6
14. Liang C (2022, in press) Technology and terror: the new arsenal of anarchy. In Masys AJ (ed) Handbook of security science. Springer Publishing. https://www.mcafee.com/enterprise/en-us/about/newsroom/press-releases/2021/20211109-01.html

15. Masys AJ (2012) Black swans to grey swans—revealing the uncertainty. Int J Disaster Prev Manag 21(3):320–335
16. Masys AJ (2016) Counter-terrorism and design thinking: supporting strategic insights and influencing operations. In: Masys AJ (ed) Disaster forensics: understanding root cause and complex causality. Springer Publishing
17. Masys AJ (2021) The security landscape—**systemic risks** shaping non-traditional security. In: Masys AJ (ed) Sensemaking in security. Springer Publishing
18. Masys AJ (2022, in press) Examining systemic risk in the cyber landscape. In: Adib Farhadi A, Sanders RP, Masys A (eds) The great power competition. Volume 3: Cyberspace: the fifth domain. Springer
19. Masys AJ (ed) (2016) Applications of systems thinking and soft operations research in managing complexity. Springer Publishing
20. OECD (2019) Strategic foresight for better policies strategic foresight, October 2019
21. Pourdehnad J, Wexler ER, Wilson DV (2011) Systems & design thinking: a conceptual framework for their integration. Working Paper #11–03, University of Pennsylvania. http://reposi tory.upenn.edu/cgi/viewcontent.cgi?article=1009&context=od_working_papers
22. Reez N (2019) On the development of strategy formation through strategic foresight. Example storytelling. In: Hellmann G, Jacobs D (eds) The German White paper 2016 and the challenge of crafting security studies. The Aspen Institute Deutschland E.V., Berlin, pp 40–47
23. Reez N (2021) Foresight-based leadership. Decision making in a growing AI environment. In: Jacobs G et al (eds) International security management: new solutions to complexity. Springer
24. Rossel P (2009) Weak signals as a flexible framing space for enhanced management and decision making. Technol Anal Strateg Manag 21(3):307–320
25. Senge P (1990) The fifth discipline: the art and practice of the learning organization. Doubleday Currency, New York
26. Senge P (2006) The fifth discipline: the art and practice of the learning organization. Doubleday Currency, New York
27. Taleb N (2007) The black swan: the impact of the highly improbable. Random House Trade Paperbacks, New York
28. Tuomi I (2013) Next-generation foresight in anticipatory organizations. Paper presented at the European Forum on Forward-looking Activities (EFLA), European Commission
29. Tõnurist P, Hanson A (2020) Anticipatory innovation governance: shaping the future through proactive policy making. In: OECD working papers on public governance, no. 44. OECD Publishing, Paris. https://doi.org/10.1787/cce14d80-en
30. UNDRR & UNU-EHS (2022) Understanding and managing cascading and systemic risks: lessons from COVID-19. UNDRR, Geneva; UNU-EHS, Bonn. https://www.undrr.org/public ation/understanding-and-managing-cascading-and-systemic-risks-lessons-covid-19
31. WEF (2022) Global risk report. https://www.weforum.org/reports/global-risks-report-2022
32. Wilner A (2020) Cyber futures: a preliminary scanning and foresight report. https://www. alexwilner.ca/other-publications/2020/10/30/cyber-futures-a-preliminary-scanning-and-foresi ght-report
33. Woods DD (2006) How to design a safety organization: test case for resilience engineering. In: Hollnagel E, Woods DD, Leveson N (eds) Resilience engineering: concepts and precepts. Ashgate Publishing, Hampshire
34. Wright A (2005) The role of scenarios as prospective sensemaking devices. Manag Decis 4(1):86–101

The Use of Counter Narratives to Combat Violent Extremism Online

Joseph Rees and Reza Montasari

Abstract Due to recent rises in extremism across the globe (Dean et al. in J Polic Intell Count Terror 11:121–142, 2016; Le Roux in Responding to the rise in violent extremism in the Sahel. Africa Center For Strategic Studies, 2019, p. 26) and (Jones in Int Secur 32:7–40, 2008), governments and law enforcement organisations, such as the police, have looked to new strategies to counter violent extremism (Russell and Theodosiou in Counter-extremism: a decade on from 7/7. Quilliam Foundation, 2015). Specifically, there has been an expanse of the field now widely known as Countering Violent Extremism (CVE). CVE is a highly contested area; however, upon conducting a literature review, Inserra (Revisiting efforts to counter violent extremism: leadership needed. The Heritage Foundation, 2015, p. 2) helpfully reduced the term CVE down to descriptions of interventions intended to "*stop individuals from radicalizing*". LaFree and Freilich (Annu Rev Criminol 2:383–404, 2019) distinguish counter-terrorism from CVE, they describe counter-terrorism as military responses ('hard') as opposed to non-military responses ('soft'), referred to as countering violent extremism strategies. This chapter contends that, although counter narratives tend to lack academic standing, there does appear to be a widespread acceptance that narratives influence individuals' beliefs. Therefore, in the absence of other more effective methods of CVE online, it is argued that counter-narratives (CN) should be used in informed ways by organisations such as the police (Monaghan in Crime Media Cult 18(1):21–39, 2020). This chapter will compare academic understandings of narrative and communication alongside examples of counter narratives issued against them. It will then assess the impact of such strategies and potential alternatives for CN.

J. Rees (✉)
London, UK
e-mail: joerees2007@yahoo.co.uk

R. Montasari
Department of Criminology, Sociology and Social Policy, School of Social Sciences, Swansea University, Swansea, Wales, UK
e-mail: Reza.Montasari@Swansea.ac.uk
URL: http://www.swansea.ac.uk

© The Author(s), under exclusive license to Springer Nature Switzerland AG 2023
R. Montasari et al. (eds.), *Digital Transformation in Policing: The Promise, Perils and Solutions*, Advanced Sciences and Technologies for Security Applications, https://doi.org/10.1007/978-3-031-09691-4_2

Keywords Extremism · Violent extremism · Counter narratives · Terrorism · Radicalisation · The Internet · Cyber terrorism · Digital policing · Countering violent extremism online · Online radicalisation

1 Introduction

Due to recent rises in extremism across the globe [28, 48, 83], governments and organisations have looked to new strategies beyond those previously used in order to counter violent extremism [85]. In the era of digital policing, a plethora of methods have been used globally by police and government agencies in order to counter criminal activity. In recent years, various examples in digital policing have been discussed in mainstream media, in particular the use mass surveillance programs such as the NSA revelations [6]. Yet, digital policing also encompasses processes such as digital forensics and even covert online investigations [32] and academics and organisations frequently discuss the development of new processes. Further, although nation states and police departments have strived to heighten their surveillance abilities, technological developments are undoubtedly presenting difficulties for law enforcement officials and organisations [55].

This chapter is premised on the understanding that extremists now have the potential to communicate with individuals across the globe like never before [36]. For example, during 2014, ISIS launched a large scale Twitter campaign in praise of the ISIS caliphate. The campaign was able to circumvent various censorship methods online, by example utilising software such as *"The Dawn of Glad Tidings"* mobile application [65]. From a digital policing perspective, Bodine-Baron et al. [16] found that, after geospatial analysis of tweets supporting ISIS, many individuals were geographically distanced, and outside of jurisdiction of governments in the West. Further, at the time of the study, the team noted 4.5 million Tweets in support of ISIS, thus suggesting that the sheer scale of online posts may be too great for organisation such as the EU Internet Referral Unit (EU IRU) to handle. Therefore, with the circumventing of internal moderation processes by platforms, and the lack of resources of other organisations to report such content, the appeal of counter-narratives is elevated. Further, counter-narratives provide a less invasive method of countering violent extremism, without the involvement of contentious issues such as the growing concerns regarding mass surveillance, and the right to privacy [103].

Correspondingly, there has been an expanse of the field now widely known as Countering Violent Extremism (CVE). CVE is a highly contested area. However upon conducting a literature review, Inserra [45, p. 2] reduced the term down to "stop individuals from radicalizing". LaFree and Freilich [54] distinguish counter-terrorism from CVE, in which they describe counter-terrorism as military responses ('hard') as opposed to non-military responses ('soft'), referred to as countering violent extremism strategies. This chapter will argue that, although counter narratives do tend to lack academic standing, there does appear to be a widespread acceptance that narratives influence individuals' beliefs. Therefore, in the absence of other more effective methods of CVE online, it is argued that counter-narratives (CN) should continue to be used. This chapter will compare academic understandings of narrative and communication alongside examples of counter narratives issued against them. It will then assess the impact of such strategies and potential alternatives for CN.

2 Extremism

When creating CVE strategies, difficulties arise when discussing violent extremism, primarily with the definition of extremism itself. For example, while there have been calls for a uniform definition [34] in order to aid policy makers, scholars such as Williford [106] caution against rushing to form a definition as this action may place limitations on free speech and lead to mislabelling certain groups as extremists. Berger [10, p. 44] refers to extremism as "*the belief that an in-group's success or survival can never be separated from the need for hostile action against an out-group*". Yet, it should be noted that although Berger [10, p. 23] provided a definition, he emphasises that the concept of "extremism is rarely simple" in that it is not particular to any singular school of politics, religion or race. Governments such as the UK government have also attempted to define the term. The UK government's 'summary definition' of extremism is: "*beliefs and actions that fall outside of mainstream or moderate values*" [100, p. 6]. Interestingly, within the same government report outlining the definition, the author felt it necessary to contain an additional definition, drawn from social psychology, which describes extremism as: "*a response to threats (perceived or otherwise) to an identified in-group*" (p. 7). The very presence of two definitions within the report further emphasises the difficulties in producing a singular definition of extremism even at a governmental level. These difficulties inevitably manifest themselves in the related discussions on the subject of violent extremism.

2.1 *Violent Extremism*

Berger [10, p. 45] states that violent extremism "*is the belief that an in-group's success or survival can never be separated from the need for violent action against an out-group*". He then goes on to say a violent extremist ideology may "*characterize its violence as defensive, offensive or pre-emptive*" [10, p. 46]. Neumann [70] notes that the concept of 'violent extremism' is viewed as behavioural as opposed to idealistic. Thus, one can be an idealistic 'extremist' without being a violent extremist. Recent rises in extremism, alongside policy and academic uncertainty have "*given rise to a wide range of charities, governments, activists, and think-tanks aiming to get involved in the field that has become known as countering violent extremism*" [57, p. 68]. Contemporary academia is assessing the role in which extremist narratives play in individuals moving to violent extremism, particularly online [41].

2.2 Aims of Extremist Narratives

In order to assess counter-narratives, the impact of the initial narrative distributed by violent extremist groups must first be analysed. Although again there is much debate surrounding the definition of narratives, some academics believe that extremist groups utilise these 'vehicles of persuasion' [22] "*to convey ideology, values, justifications, or core concerns to sympathizers, would-be members, and the greater public*" [21, p. 1]. Quiggins [77, p. 23] summarises extremist narratives as possessing three parts: the beginning which explains a grievance or difficulty, followed by a middle containing hero, agent or potential solution to the problem and an end which provides a solution and/or challenge to the audience, to act for themselves. Berger [10] has similar findings in what he describes as a 'crisis-solution construct'; in which extremist movements use in-group crisis' in order to justify their extraordinary 'solutions'. These narratives play to social identity theory which stipulates that individuals place themselves in competing social groups, resulting in the formation of in-group and out-group identification [61]. The Linkage Based Model [44] theorises this concept, by explaining that an interplay (linkage) between the in-group out-group and the previously mentioned crisis-solution concept produces an overarching narrative which in turn produces "*cyclically self-reinforcing narratives*" [42, 43, 78] and creating a core narrative.

3 Narratives

3.1 Online Narratives

Whilst researching various datasets including violent extremists within the U.K., Gill et al. [35, p. 113] concurred that although not essential for violent radicalisation and attack planning, the internet is indeed a "facilitative tool". It is important to note that although the internet does provide opportunities for surveillance and strategic communications, including counter-narratives [4], in recent years media speculation and academic studies have been focused on extremists use of the internet. In particular, part of a noted shift from offline radicalisation to the study of individuals who have been potentially radicalised online as a result of exposure to extremist narratives [51]. For example, in 2007, when referring to Al-Qaeda's use of technology to post online narratives, US Defence Secretary Robert Gates was quoted saying: "*How has one man in a cave managed to out-communicate the world's greatest communication society?*" (c.f. [68, p. 134]). This statement by a senior politician in the USA emphasises the transition by violent extremists into the use of online narratives and their potential power, reach and significance in modern society [82].

Another contemporary example of potential online radicalisation relates to the Islamic State of Syria and Levant (ISIS) and the unprecedented global recruitment of its members [62]. While recognising the vulnerability of youth to extremist narratives

[91], reducing the reasoning behind the radicalisation to ISIS down to "rebellious Muslim youths" [89, p. 1] perhaps distracts from other variables at play. Specifically, it is emphasised online narratives may well have contributed to the radicalisation and subsequent recruitment of ISIS members, due to the potential of these narratives to *"inspire people to do terrible things, or to push back against those extremist voices"* [40, p. 2]. In essence, it comes as no surprise that extremist groups are utilising available technology in order to further their cause as has been seen in the past [8]. It should not be assumed, however, that violent extremists' use of the internet is a recurrence of earlier academic concerns such as Lippmann's [60] fears for the influential effect of newspapers. Such an assumption may represent an underestimation of the influencing power and global reach of the internet [24, 104]. It is well documented that the internet has been used by violent extremist groups for a multitude of reasons from financing to communication [46, 47] and, at the very least, as a "facilitator" of violent extremism [37, p. 2]. Studies of this nature, when combined with the prominent manifestation of extremist narratives online (including, for example, the ISIS 2014 Twitter campaign—see [16]) have resulted in calls for further 'countermeasures' to extremist online propaganda [33].

3.2   Narrative Persuasion

Although some have argued that ideological justification of violence may involve motives instead of or in addition to ideology [93], many governments have assumed that violent extremism is ideologically based and therefore have promoted CVE strategies that help to keep individuals from being radicalised [105]. Berger [10, p. 46] defines radicalisation into extremism as a spectrum and not a 'singular destination': *"an escalation of an in-group's extremist orientation in the form of increasingly negative views about an out-group or the endorsement of increasingly hostile or violent actions against an out-group"*. This 'persuasion into radicalisation' has been frequently discussed (see [23, 81]) and more specifically, the use of "weaponized words" within this process [20, p. 42]. In order to create policies and implement counter strategies to these potentially persuasive narratives, the role of narratives in persuading a target audience must be assessed ([3 c.f.]). This assessment addresses issues such as whether the same narrative persuasion can be used in the production of counter-narratives to achieve a key objective of CVE, that is, dissuading support for terrorism [63].

3.3   Call for Counter Narratives

At a rapidly growing rate, counter-narrative strategies are now present in many CVE policies around the world, including those issued by countries, such as the US (Global Engagement Centre), UK (Government and Global Coalition) and by organisations such as the UN [102]. Some have linked the push for further 'soft' CVE policies to previous conflicts, namely the US War on Terror [11]. The heavily publicised "credibility gap" [19, p. 16] between US policies and actions led to much criticism by media and politicians, in what one academic described as the 'Say-do-gap' [80, p. 271]. In a pressured search of new CVE strategies, counter-narratives certainly fall within the bounds of what English [29, p. 133] refers to as the 'best responses' to violent extremism—*"those that respect legal frameworks and the democratically established rule of law"*. It was thought that policies could be created that would prevent radicalisation into violent extremism and also represent a positive intervention directed towards those already in the process of radicalisation. Whether this remains the reasoning behind the incorporation of counter-narratives into CVE methods is unclear; what is the evident, is the widespread implementation of counter-narrative strategies into CVE policies. Alternative CVE methods involving the removal of 'harmful' content online have led to concerns of censorship and the inability to express free speech online [10] by both 'regular citizens' and extremists alike. As previously described, the very nature of extremism is beliefs outside of mainstream viewpoints. Thus, the removal of online material may appear to be an attempt, for politically-driven purposes, to undermine valuable narratives [56]. This contentious issue has led to difficulties in previous strategies and is still a difficulty when strategising counter violent extremism, particularly in locations such as the United States of America where free speech is protected under the constitution, reaffirmed in cases such as *Brandenburg v Ohio* [97]. Arguably, legal cases such as these, may explain the promotion of counter-narratives as a counter-extremism strategy in the U.S.A.

Hence, while counter-narratives will indeed be affected by politics, as witnessed with the Trump administration's funding changes [5], they do provide a sense of audience choice and refrain from censorship. Some of the most prominent examples of counter-narratives can be seen in response to online posts by ISIS. For example, specifically in relation to ISIS, Speckhard et al. [94] state that police forces in countries such as Kyrgyzstan, Holland and Belgium have used counter-narrative videos, both online and in training events, one of the notable features of these videos is that they are given pro-ISIS titles and begin with material actually taken from ISIS propaganda material. At the height of ISIS social media postings, Zeiger and Aly [108, p. 73] stated that *"In order to break the present momentum of ISIS, the use of military force in Iraq and Syria is not enough. What is needed is a counter-narrative that can seriously challenge and undermine the narrative of ISIS"*. Although there were calls from the academic community to incorporate this strategy, there was also resistance from academics who questioned the underlying theories to the strategy. One such academic, Szamnia and Fincher [98] emphasises the necessity for policy makers to utilise research findings during the development of CVE policies and programs. Gill et al. [35] reinforces this approach, even if it means challenging long held assumptions in the field of CVE. Some critics go further and state that the policy of counter narratives simply provides 'oxygen' for extremists [21], which if correct suggests that counter-narratives have little role to play in CVE online. In challenging the basis of counter narratives, this view again calls into question what is meant by the very term counter narrative. This issue is considered in more detail below.

3.4   Counter Narratives Defined

Hemmingsen and Castro [38, p. 9] describe the term 'counter narratives' as "*a broad and ill-defined concept that is used in very different ways*". Neumann [71, p. 433] discusses definitional issues by stating that counter narratives work: "*by discrediting, countering, and confronting extremist narratives or by educating young people to question the messages they see online in order to attempt to reduce the demand for extremist narratives*". To bring further clarity, Briggs and Feve [19] differentiate counter-narratives from alternative narratives by stating that counter-narratives are used to discredit the extremist narrative; in contrast, alternative-narratives publish a different positive narrative. Adding further difficulty to producing a definition of CN, the aims of counter-narratives are not agreed either. From an ideological perspective the aim of a counter narrative is to "*plant a seed of question*" in the hope of an individual "questioning their group ideology" [13]. This aim is often viewed as both overly broad and ambitious; thus, Leuprecht et al. [59, p. 33] concisely defines the counter-narrative task as one "*to counter those narratives with the clearest link to violence*". Schmid [89, p. 3] states that counter-narratives are aimed at those who are "*not yet fully radicalised*". In order to assess the potential of persuasion of both narratives and counter narratives, the Elaboration Likelihood Model of Persuasion (ELMP) [76] has been utilised in this chapter. This ELMP model is an appropriate choice for the analysis of both narratives and counter-narratives as their innate goal is persuasion and its applicability to the study of assessing shifts or changes in attitude is accepted by many within the behavioural community [92]. In addition, previous studies have applied the ELMP model to relevant areas including propaganda [67], online hate speech [58] and violent extremist narratives [33]. In addition to the ELMP model, two counter narrative case studies will be used in order to assess their effectiveness in the dissuasion away from violent extremism namely the 'Think Again Turn Away' (TATA) campaign and the 'ExitUSA' campaign.

T.A.T.A campaign was established by the center for Strategic Counterterrorism Communications, targeting propaganda of the Islamic State online, by Tweeting and in particular responding directly with ISIS Twitter accounts [42, 43]. In comparison, the ExitUSA campaign was run by 'Life After Hate' a non-profit organisation which targeted far right extremists within the USA and offer 'a way out' [53] by increasing awareness of far right extremism in the us (primarily using a Twitter page) and by interacting directly with extremists online.

3.5   Elaboration Likelihood Model of Persuasion

The 'Elaboration Likelihood Model of Persuasion' (ELMP) is based in the premise that the modification of attitudes can take place with a high degree or a low degree of thought [58]. The ELMP suggests that this occurs as part of a dual process of message elaboration which is due to both their motivation and ability [76]. The peripheral processing route (P.P.R) is where low end processing takes place while the central processing route (C.P.R) is where high end processing takes place. Although it is possible to perform low and high end processing independently, the model indicates that persuasion is frequently a combination of these processes [33]. The degree of thought determines how consequential the judgement is, with the P.P.R usually producing 'ephemeral attitudes' and the C.P.R producing more stable long term effects [75]. Attempting to establish primary use of the central processing route is advantageous as: *"central route attitude changes are usually longer lasting and more predictive of behavioural changes, when compared to the peripheral route"* [64, p. 34]. A literature review of studies on hate crime and propaganda using the ELMP by Schieb and Preuss [87, p. 588] identified three key predictors of the likelihood of persuasion from the model, namely: **target characteristics, message characteristics** and **source characteristics**. Notably, in a recent literature review, Hamid [39] similarly found these effectors to be key in mass communications persuasiveness. In the discussion below, these three predictors are considered with specific reference to narratives and counter narratives.

4   Characteristics

4.1   Target Characteristics

Earlier theories of mass communication such as the 'hypodermic needle theory' in which it was believed an audience could be 'injected' with a narrative have slowly been overtaken [14]. Narratives are now understood to be interpreted differently by different individuals [39], affected by a culmination of factors referred to here as target characteristics. Yet just as online extremists narratives cannot easily be targeted towards a particular group due to online anonymity [10], there are also difficulties in targeting a group with counter narratives. Although, it should be noted key words relating to ISIS and redirected them to pre-curated anti-ISIS YouTube videos [84]. However, other counter-narratives have been accused of disseminating their messages in the 'hope' of them reaching their target audience. Bélanger et al. [7] suggest that 'narrowcasting' their message to their target audience is a more optimal means of dissemination as opposed to traditional broadcasting.

The need for cognition is another important aspect of the target characteristics, as individuals who have a high need for cognition and knowledge expansion are more likely to be affected by persuasive messaging [87]. Therefore, during what Berger

[10, p. 125] refers to as 'the curiosity about the extremist in-group' stage of radicalisation, there is significance as to the 'curiosity' of the individual in the persuasiveness of the process. This perhaps implies that if there is indeed counter evidence in regard to the in-group, the curiosity of the individual could be turned elsewhere. Perhaps those within the early stages of radicalisation (seeking information) may also be more susceptible to information from a counter narrative. This would suggest that counter-narratives would be particularly useful for those on the path to radicalisation as a means to counter violent extremism online, but may be limited when attempting to address those more attuned to extremist narratives. Similar findings have been found in consumer behaviour. Thomas [99] found that advertising was more effective to consumers who were subject to advertising during a consumer uncertainty period. However it should be noted that Hamid [39] argues that individuals form non-negotiable values, 'sacred values' show more resistance to change than non-sacred values, thereby potentially limiting the applicability of consumer behaviour studies to extremism.

The individual's motivation when processing a message is also indicative of the likelihood of persuasiveness of the message. Petty and Cacioppo [76, p. 121] stated that an increase in individual motivation can act "as argument scrutiny is increased, peripheral cues become relatively less important determinants of persuasion". Therefore, the crisis-solution construct that is exploited by extremists groups potentially increases the targets' motivation, therefore impacting persuasiveness [88]. The TATA campaign appeared to attempt this kind of offensive discrediting tactic. For example, one tweet asked: "What is the difference between members of DAESH and hyenas?". Attached was a picture of a pack of hyenas killing its prey, and a video of ISIS members beheading a victim [18]. As Stevens and Neumann [96] suggested, a reduction in motivation through discrediting and removing relevance of an extremist group is the most effective, as it could refute claims of a fabricated crisis perhaps the individual targets motivation will be lowered, in turn reducing the persuasiveness of the extremist message. This view strongly suggests that online counter narratives have a role to play in countering extremism, primarily as a means to discredit extremist crisis construction.

4.2   Message Characteristics

As previously mentioned, Hamid [39] conducted a literature review in which sacred values among extremists were identified as non-negotiable, further, these values were linked to some groups' willingness to make personal sacrifices in defence of their values when these values where seemingly under attack [39, 90]. This is understood within the ELMP as *"Variables affecting message processing in a relatively biased manner can produce either a positive (favorable) or negative (unfavorable) motivation"* [76, p. 127]. Similarly neuroscience research has found that opposing political viewpoints can trigger defensiveness [49]. Similarly, upon research Bélanger et al. [7]

concluded that 'backfire' was possible and in fact consistent when challenging religion in particular. A further risk of offensive counter narratives (as opposed to defensive) is that they can easily escalate. Katz [50] suggested that the TATA campaign on twitter is "not only ineffective, but also provides jihadists with a stage to voice their arguments". When viewing more general political rhetoric online, Kovaleva [52] found that even states with vast resources such as the Russian government, have reduced narrative controls when online, as opposed to offline. Within early messages of the TATA campaign, the lack of narrative control the government agency possessed online was evident. As noted by Katz [50] in a media article, this was evident in the very occurrence of the TATA campaign discussing the Abu Ghraib prison in which US troops mistreated prisoners in Iraq, a topic of contention in the US [9]. At the very least, the discussion surrounding sacred values and counter narratives, offensive in nature, calls for great care in the framing and expression of counter narratives if they are to be effective. For example, Hamid [39] has suggested that, to minimise defensiveness and backfire, counter narratives which challenge the social norms regarding violence are likely to be more effective than counter narratives which challenge beliefs directly.

4.3   Source Characteristics

In recent years, the organisation and sophistication of online campaigns by groups such as ISIS have created many difficulties for governments. The complexity of the infrastructure within ISIS territory, alongside the sophistication of their online campaigns created a sense of credibility for their message which, in turn, made it more persuasive [69]. Interestingly, both extremist narratives and counter-narratives have employed 'champions' as a means of increasing the source credibility and receptiveness of the message [46, 47]. In the case of counter-narratives, these champions may even include former radicals as in the case of ExitUSA [53]. Indeed, the source factors of persuasion are very important in the acceptance of a message [76, p. 124]. The existence of a 'Say-do-gap' between government policy and narratives reduces trust in the credibility of the message source in turn reducing its persuasive ability [1]. Both the TATA and ExitUSA counter-narratives faced problems of credibility as both were either directly involved or partially funded by government bodies. Interestingly, as previously mentioned Jigsaw a different counter narrative programme used pre-existing content as a source of counter-narrative material. From the literature it appears that, although beliefs can be strengthened by means of mass communication, belief formation is taking place elsewhere. Schils and Verhage [86] argue that belief formation is taking place largely offline in the physical environment. Notably, Papadopouloset et al. [73] suggest that online sites (including social media) do in fact have components of offline environments, thus the studies of persuasive impact of these offline groups could also be applicable to online cases. Yet when reviewing individuals who joined Al-Qaeda and ISIS, Perliger and Milton ([74], c.f.

[39]) found that, of those studied, there was considerable person-to-person inter-action (offline or online). In addition the assumption that online communities are highly similar to offline communities fails to explain the geographical clustering of extremist individuals as seen in studies of individuals from Western countries moving to Iraq and Syria to join ISIS [17]. Further the very existence of priority areas within the United Kingdom's Prevent Strategy [17] suggests that the existence of offline networks is, perhaps, a larger contributor to violent extremism, than the mass communication of narratives online. The ELMP also accounts for this, poten-tially downplaying the effectiveness of counter narratives, but also the original violent extremist narratives online, leaving unresolved conclusions for the effectiveness of the use of counter-narratives as a means of countering violent extremism.

5 Assessing the Success of Counter Narratives

The assessment of 'hard' approaches to violent extremism, where the rationalist approach is commonly applied to military perspectives [95], in which the number of threats neutralised is counted, cannot be used directly with counter-narratives. The assumption of the use of counter-narratives online would be the availability of large scale metrics. Yet, notably counter narrative campaigns resort to much more general measures of success. In the case of Harakat-ut-Taleem a counter narrative against jihadist propaganda online, three metrics to measure the success—awareness, engagement and impact [91]. Primarily, the difficulty in the assessment of counter-narratives is the inability to measure belief and/or attitude changes. If engaging with extremists was the aim of these campaigns, they would certainly be classed as a success. Both ExitUSA and TATA campaigns actively communicated with extrem-ists, with TATA in particular being called a "Twitter War" [50]. The measurement of belief change either explicitly (direct questioning) or implicit measurements (online engagements etc.) would both require knowledge of which members of an online community were actually of the target audience [105]. In addition even if the target audience could be identified through the anonymity offered by the internet, there would then be difficulty assessing its impact on a belief system, not just general engagement. For example, Bull and Rane [25] raise the possibility that certain counter narratives may have unintended negative consequences such as the social marginal-isation of certain communities and individuals. Without a 'baseline' of belief and perceptions in many campaigns [31], attitudinal changes would be difficult to estab-lish as engagement only offers a momentary indication of a viewpoint [91]. This analysis is further complicated due to the anonymity online which not only provide difficulties for follow up but also creates further problems enhanced in online envi-ronments [10]. This includes the potential for individuals' dishonest 'posturing' of extremist views online which they do not actually hold offline [39]. One study found that individuals subject to opposing views did in fact object with anger at the time (engaging emotionally online), but later questioned their beliefs, which would be extremely difficult to measure [91].

6 Alternatives to Counter Narratives

Calls for internet regulation are not new, though there have been relatively recent calls for the regulation of social media platforms which "have inarguably become the very tools that aid in ISIS incitement and recruitment" [107, p. 45] and have been at the forefront of the discussion. As with previous calls for the regulation of the internet, these calls were met with opposition, due to the previously mentioned freedom of speech implications. When looking for alternative CVE strategies on platforms such as social media there are further difficulties beyond free speech. In the UK the recent proposal for the 'Online Harms White Paper' [101] by the UK government towards the removal of 'online harms', including for example terrorist related content, outlined the difficulties in the process of removal both technologically and legally [15]. Similarly, Europol's Internet Referral [30] unit identifies violent extremist content online and then contacts hosts of material asking for the removal of the content as it is conflicts with the hosts own terms of service. When compared with the use of counter narratives the above methods are much more resource intensive, from a monetary aspect alone, the entire TATA campaign cost $5M compared to the vast expense of running Europols Internet Referral Unit. Due to the now immense quantities of violent extremist content online, all of the content cannot be taken down nor can every message be rebutted with a counter-narrative. However, at the very least, showing some presence on these platforms is important when fighting violent extremist ideology [12]. In addition, arguments for further moderation of online spaces simply force the migration of extremist content to smaller platforms with less resources, in turn creating unmoderated 'echo chambers' for extremist ideas [72], areas of 'repeated narrative', a factor in message persuasion [87, p. 584]. Although this kind of moderation will reduce the reach of extremist groups, the concept of unopposed in-group narratives creates environments for persuasion, and in turn radicalisation.

7 Conclusion

Although recent studies have concluded that people are indeed less vulnerable to mass persuasion than previously thought [39], mass communication theory remains influential in the lives of individuals throughout the world. Perhaps, as with the advertising industry, the promotion of counter narratives in recent years has been subject to prestige bias [2] in which larger more powerful countries and organisations have promoted the practise pressuring other countries to follow. Although not as "inherently intuitive" [79, p. 1] as it first appears, counter-narratives do offer another tool to fight violent extremism online. Bélanger et al. [7] argue that *"counter-narratives are the cornerstone short-term intervention in the fight against violent extremism"*. Nevertheless, what is likely needed is a multifaceted approach to CVE online including counter-narratives and working alongside tech companies to remove extreme content online. The ideological war of ideas is likely to remain for years to come, but if countries can indeed decrease the 'credibility gap' it may promote a higher degree of trust towards both alternative and counter-narratives and in turn help the fight against violent extremism [26]. Although some 'harder' approaches to countering violent extremism deal with immediate threats, counter-narratives attempt to fight an ideology and may therefore offer utility beyond a short-term nature. Combining attempts to address the root causes of violent extremism and the reduction of military intervention where possible, could decrease the credibility gap and avoid creating further grievances in turn reducing the likelihood of violent extremism [27]. It is concluded that, although counter-narratives certainly gain the attention of individuals online, and likely increase awareness, their overall impact is likely to be limited. What is needed is a combination of counter narratives with alternate narratives in order to both raise awareness and offer opportunities outside of the extremist narrative; alongside offline CVE approaches, as perhaps the assumed dichotomy between online and offline radicalisation may in fact be false [35, p. 114].

References

1. Aistrope T (2016) The Muslim paranoia narrative in counter-radicalisation policy. Crit Stud Terror 9(1):182–204. https://doi.org/10.1080/17539153.2016.1175272
2. Ackerberg DA (2001) Empirically distinguishing informative and prestige effects of advertising. RAND J Econ 32(2):316–333
3. Aly A, Weimann-Saks D, Weimann G (2014) Making "noise" online: an analysis of the say no to terror online campaign. Perspect Terror 8(5):3–47
4. Aly A, Macdonald S, Jarvis L, Chen T (2016) Violent extremism online: new perspectives on terrorism and the internet. Routledge
5. Aziz SF (2017) Losing the war of ideas: a critique of countering violent extremism programs. Texas Int Law J 52:255
6. Bauman Z, Bigo D, Esteves P, Guild E, Jabri V, Lyon D, Walker RB (2014) After snowden: rethinking the impact of surveillance. Int Polit Sociol 8(2):121–144
7. Bélanger JJ, Nisa CF, Schumpe BM, Gurmu T, Williams MJ, Putra IE (2020) Do counter-narratives reduce support for ISIS? Yes, but not for their target audience. Front Psychol 11(1):126–144
8. Benson DC (2014) Why the Internet is not increasing terrorism. Secur Stud 23(2):293–328
9. Bennett WL, Lawrence RG, Livingston S (2006) None dare call it torture: indexing and the limits of press independence in the Abu Ghraib scandal. J Commun 56(3):467–485
10. Berger JM (2018) Extremism. MIT Press
11. Berger JM (2016) Making CVE work. International Centre for Counter-Terrorism, The Hague, p 7
12. Bertram L (2016) Terrorism, the Internet and the social media advantage: exploring how terrorist organizations exploit aspects of the Internet, social media and how these same platforms could be used to counter-violent extremism. J Deradical 7:225–252
13. Bertram L (2015) How could a terrorist be de-radicalised? J Deradical 23(1):44–63
14. Bineham JL (1988) A historical account of the hypodermic model in mass communication. Commun Monogr 55(3):230–246
15. Bishop P, Macdonald S (2019) Terrorist content and the social media ecosystem: the role of regulation. Digital Jihad: Online Commun Viol Extrem 31(1):135–152

16. Bodine-Baron E, Helmus TC, Magnuson M, Winkelman Z (2016) Examining ISIS support and opposition networks on Twitter. RAND Corporation
17. Bouhana N, Marchment Z, Schumann S (2018) Ours is not to reason why, but where: investigating the social ecology of radicalisation. American Society of Criminology Annual Meeting, Atlanta, GA
18. Bouzis K (2015) Countering the Islamic State: US counterterrorism measures. Stud Confl Terror 38(10):885–897
19. Briggs R, Feve S (2013) Review of programs to counter narratives of violent extremism: what works and what are the implications for government? Institute for Strategic Dialogue
20. Braddock K (2020) Weaponized words: the strategic role of persuasion in violent radicalization and counter-radicalization. Cambridge University Press
21. Braddock K, Horgan J (2016) Towards a guide for constructing and disseminating counternarratives to reduce support for terrorism. Stud Confl Terror 39(5):381–404
22. Braddock K, Dillard JP (2016) Meta-analytic evidence for the persuasive effect of narratives on beliefs, attitudes, intentions, and behaviors. Commun Monogr 83(4):446–467
23. Braddock KH (2012) Fighting words: the persuasive effect of online extremist narratives on the radicalization process. Pennsylvania State University Press
24. Brown K, Pearson E (2018) Social media, the online environment and terrorism. In: Routledge handbook of terrorism and counterterrorism. Routledge
25. Bull M, Rane H (2019) Beyond faith: social marginalisation and the prevention of radicalisation among young Muslim Australians. Crit Stud Terror 12(2):273–297
26. Cherney A, Murphy K (2017) Police and community cooperation in counterterrorism: evidence and insights from Australia. Stud Confl Terror 40(12):1023–1037
27. Crenshaw M (1981) The causes of terrorism. Comp Polit 13(4):379–399
28. Dean G, Bell P, Vakhitova Z (2016) Right-wing extremism in Australia: the rise of the new radical right. J Polic Intell Count Terror 11(2):121–142
29. English R (2010) Terrorism: how to respond. Oxford University Press
30. Europol (2019) 2019 EU Internet referral unit transparency report. https://www.europol.europa.eu/publications-documents/eu-iru-transparency-report-2019. Accessed April 2021
31. Europa (2017) Countering terrorist narratives. Policy Department for Citizens Rights and Constitutional Affairs. https://www.europarl.europa.eu/RegData/etudes/STUD/2017/596829/IPOL_STU(2017)596829_EN.pdf. Accessed April 2021
32. Fussey P, Sandhu A (2020) Surveillance arbitration in the era of digital policing. Theor Criminol 26(1):3–22
33. Frischlich L, Rieger D, Morten A, Bente G (2018) The power of a good story: narrative persuasion in extremist propaganda and videos against violent extremism. Int J Confl Viol (IJCV) 12(1):42–63
34. Gelfand MJ, LaFree G, Fahey S, Feinberg E (2013) Culture and extremism. J Soc Issues 69(3):495–517
35. Gill P, Corner E, Conway M, Thornton A, Bloom M, Horgan J (2017) Terrorist use of the Internet by the numbers: quantifying behaviors, patterns, and processes. Criminol Public Policy 16(1):99–117
36. Ganesh B, Bright J (2020) Countering extremists on social media: challenges for strategic communication and content moderation. Policy Internet 12(1):6–19
37. Gaudette T, Scrivens R, Venkatesh V (2020) The role of the Internet in facilitating violent extremism: insights from former right-wing extremists. Terror Polit Viol 21(1):1–18
38. Hemmingsen AS, Castro KI (2017) The trouble with counter-narratives (No. 2017: 01). DIIS report. https://www.econstor.eu/bitstream/10419/197640/1/880712619.pdf. Accessed March 2021
39. Hamid N (2020) The ecology of extremists' communications: messaging effectiveness, social environments and individual attributes. RUSI J 165(1):54–63
40. Harman J (2014) Future terrorists. Los Angeles Times. http://articles.latimes.com/2014/jan/06/opinion/la-oe-harman-terrorism-response20140106. Accessed March 2021

41. Hardy K (2018) Comparing theories of radicalisation with countering violent extremism policy. J Deradical 15:76–110
42. Ingram HJ (2016) A brief history of propaganda during conflict. The International Center for Counter-Terrorism, The Hague, p 7
43. Ingram HJ (2016) Deciphering the siren call of militant Islamist propaganda: meaning, credibility & behavioural change. International Centre for Counter-Terrorism, The Hague
44. Ingram HJ (2017) The strategic logic of the "linkage-based" approach to combating militant Islamist propaganda: conceptual and empirical. International Centre for Counter-Terrorism, The Hague
45. Inserra D (2015) Revisiting efforts to counter violent extremism: leadership needed. The Heritage Foundation, No. 4390
46. Jacobson M (2010) Terrorist dropouts: learning from those who have left. Washington Institute for Near East Policy
47. Jacobson M (2010) Terrorist financing and the Internet. Stud Confl Terror 33(4):353–363
48. Jones SG (2008) The rise of Afghanistan's insurgency: state failure and Jihad. Int Secur 32(4):7–40
49. Kaplan JT, Gimbel SI, Harris S (2016) Neural correlates of maintaining one's political beliefs in the face of counterevidence. Sci Rep 6(1):1–11
50. Katz R (2014) The State Department's Twitter war with ISIS is embarrassing. Time Magazine. https://time.com/3387065/isis-twitter-war-state-department. Accessed March 2021
51. Koehler D (2014) The radical online: individual radicalization processes and the role of the Internet. J Deradical 1:116–134
52. Kovaleva N (2018) Russian information space, Russian scholarship, and Kremlin controls. Def Strat Commun 4(4):133–171
53. Life After Hate (2020) About us page. https://www.lifeafterhate.org/about-us-page. Accessed April 2021
54. LaFree G, Freilich JD (2019) Government policies for counteracting violent extremism. Annu Rev Criminol 2(1):383–404
55. Lavorgna A (2015) Organised crime goes online: realities and challenges. J Money Laund Control 18(2):153–168
56. Lee B (2019) Informal counter messaging: the potential and perils of the informal counter messaging space. Stud Confl Terror 42(1):161–177
57. Lee B (2020) Countering violent extremism online: the experiences of informal counter messaging actors. Policy Internet 12(1):66–87
58. Lee E, Leets L (2002) Persuasive storytelling by hate groups online: examining its effects on adolescents. Am Behav Sci 45(6):927–957
59. Leuprecht C, Hataley T, Moskalenko S, McCauley C (2009) Winning the battle but losing the war? Narrative and counter-narratives strategy. Perspect Terror 3(2):25–35
60. Lippmann W (1922) Public opinion. New York: Macmillan
61. Mackie DM (1986) Social identification effects in group polarization. J Pers Soc Psychol 50(4):720
62. Mahood S, Rane H (2017) Islamist narratives in ISIS recruitment propaganda. J Int Commun 23(1):15–35
63. McCants W, Watts C (2012) U.S. strategy for countering violent extremism: an assessment. Foreign Policy Research Institute. http://www.fpri.org/articles/2012/12/us-strategy-countering-violent-extremism-assessment. Accessed April 2021
64. McNeill BW (1989) Reconceptualizing social influence in counseling: the Elaboration Likelihood Model. J Couns Psychol 36(1):24–33
65. Monaci S (2017) Explaining the Islamic state's online media strategy: a transmedia approach. Int J Commun 11(1):19
66. Monaghan J (2022) Performing counter-terrorism: police newsmaking and the dramaturgy of security. Crime Media Cult 18(1):21–39
67. Müller F, van Zoonen L, Hirzalla F (2014) Anti-Islam propaganda and its effects: fitna, fear-based communication and the moderating role of public debate. Middle East J Cult Commun 7(1):82–100

68. Nacos BL (2009) Revisiting the contagion hypothesis: terrorism, news coverage, and copycat attacks. Perspect Terror 3(3):3–13
69. Nashar M, Nayef H (2019) 'Cooking the meal of terror' manipulative strategies in terrorist discourse: a critical discourse analysis of ISIS statements. Terror Polit Viol 3:1–21
70. Neumann PR (2003) The trouble with radicalization. Int Aff 89(4):873–893
71. Neumann PR (2013) Options and strategies for countering online radicalization in the United States. Stud Confl Terror 36(6):431–459
72. O'Hara K, Stevens D (2015) Echo chambers and online radicalism: assessing the Internet's complicity in violent extremism. Policy Internet 7(4):401–422
73. Papadopoulos S, Kompatsiaris Y, Vakali A, Spyridonos P (2012) Community detection in social media. Data Min Knowl Disc 24(3):515–554
74. Perliger A, Milton D (2016) From cradle to grave: the lifecycle of foreign fighters in Iraq and Syria. US Military Academy-Combating Terrorism Center West Point United States
75. Petty RE, Briñol P (2011) The elaboration likelihood model. Handb Theor Soc Psychol 1:224–245
76. Petty RE, Cacioppo JT (1986) The elaboration likelihood model of persuasion. In: Communication and persuasion. Springer
77. Quiggin T (2009) Understanding al-Qaeda's ideology for counter-narrative work. Perspect Terror 3(2):18–24
78. Reed A, Dowling J (2018) The role of historical narratives in extremist propaganda. Def Strat Commun 4(1):79–104
79. Reed A (2017) IS propaganda: should we counter the narrative? International Centre for Counter-Terrorism, The Hague. https://icct.nl/publication/is-propaganda-should-we-counter-the-narrative/. Accessed April 2021
80. Reed A (2017) Counter-terrorism strategic communications: back to the future—lessons from past and present. In: Conway M, Jarvis L, Lehane O (eds) Terrorists' use of the Internet: assessment and response. Los Press, pp 269–278
81. Rocca NM (2017) Mobilization and radicalization through persuasion: manipulative techniques in ISIS' propaganda. Int Relat Dipl 5(11):660–670
82. Roselle L, Miskimmon A, O'loughlin B (2014) Strategic narrative: a new means to understand soft power. Media War Confl 7(1):70–84
83. Le Roux P (2019) Responding to the rise in violent extremism in the Sahel. Africa Center for Strategic Studies, p 26
84. Redirect Method (2019) The redirect method—a blueprint for bypassing extremism. https://redirectmethod.org/downloads/RedirectMethod-FullMethod-PDF.pdf. Accessed April 2021
85. Russell J, Theodosiou A (2015) Counter-extremism: a decade on from 7/7. Quilliam Foundation
86. Schils N, Verhage A (2017) Understanding how and why young people enter radical or violent extremist groups. Int J Confl Viol 11:473–473
87. Schieb C, Preuss M (2018) Considering the elaboration likelihood model for simulating hate and counter speech on Facebook. SCM Stud Commun Media 7(4):580–606
88. Schmid AP (2013) Radicalisation, de-radicalisation, counter-radicalisation: a conceptual discussion and literature review. International Centre for Counter-Terrorism. http://www.icct.nl/download/file/ICCT-Schmid-Radicalisation-De-Radicalisation-Counter-Radicalisation-March-2013.pdf. Accessed April 2021
89. Schmid AP (2015) Challenging the narrative of the "Islamic State." International Centre for Counter-Terrorism, The Hague, pp 1–19
90. Sheikh H, Gómez Á, Atran S (2016) Empirical evidence for the devoted actor model. Curr Anthropol 57(13):204–209
91. Silverman T, Stewart CJ, Birdwell J, Amanullah Z (2016) The impact of counter-narratives. Institute for Strategic Dialogue, pp 1–54
92. Slater MD, Rouner D (2002) Entertainment, education and elaboration likelihood: understanding the processing of narrative persuasion. Commun Theory 12(2):173–191

93. Slovic P, Finucane ML, Peters E, MacGregor DG (2004) Risk as analysis and risk as feelings: some thoughts about affect, reason, risk, and rationality. Risk Anal: Int J 24(2):311–322

94. Speckhard A, Shajkovci A, Bodo L (2018) Fighting ISIS on Facebook—breaking the ISIS brand counter-narratives project. http://www.icsve.org/research-reports/fighting-isis-onface book-breaking-the-isis-brand-counter-narratives-project/. Accessed April 2022

95. Spencer A (2006) The problems of evaluating counter-terrorism. Revista UNISCI 12:179–201

96. Stevens T, Neumann PR (2009) Countering online radicalisation: a strategy for action. The International Centre for the Study of Radicalisation and Political Violence, London. https://cst.org.uk/docs/countering_online_radicalisation1.pdf. Accessed March 2021

97. Supreme Court of the United States (1968) U.S. reports: Brandenburg v. Ohio, 395 U.S. 444. Retrieved from the Library of Congress: https://www.loc.gov/item/usrep395444/. Accessed March 2021

98. Szmania S, Fincher P (2017) Countering violent extremism online and offline. Criminol Public Policy 16:119–132

99. Thomas M (2019) Was television responsible for a new generation of smokers? J Consum Res 46(4):689–707

100. U.K. Government (2019) Public perceptions of extremism report. https://assets.publishing.service.gov.uk/government/uploads/system/uploads/attachment_data/file/835681/Public_Perceptions_of_Extremism_Interim_report_Final_.pdf. Accessed April 2021

101. U.K. Government (2019) Online Harms White Paper. https://assets.publishing.service.gov.uk/government/uploads/system/uploads/attachment_data/file/973939/Online_Harms_White_Paper_V2.pdf. Accessed April 2021

102. United Nations (2017) Security Council Resolution 2017. https://www.undocs.org/S/RES/2017%20(2011). Accessed April 2021

103. Watt E (2017) The right to privacy and the future of mass surveillance. Int J Human Rights 21(7):773–799

104. Weimann G (2007) Using the Internet for terrorist. Hypermedia seduction for terrorist recruiting 25:47

105. Williams MJ (2020) Preventing and countering violent extremism: designing and evaluating evidence-based programs. Taylor & Francis

106. Williford AC (2018) Blurred lines: what is extremism. J Law Reform 52:913–937

107. Yu J (2018) Regulation of social media platforms to curb ISIS incitement and recruitment: the need for an international framework and its free speech implications. J Glob Justice Public Policy 4(1):122–136

108. Zeiger S, Aly A (2015) Countering violent extremism: developing an evidence-base for policy and practice. Curtin University

Ethical Challenges in the Use of Digital Technologies: AI and Big Data

Vinden Wylde, Edmond Prakash, Chaminda Hewage, and Jon Platts

Abstract This chapter looks at the normative and applied ethical standpoint and application of real-world technology challenges, with regard to Big Data (BD) and Artificial Intelligence (AI) domains and use-cases. In particular, how ethical technology design and utility can aid government policy makers, senior-management, software developers, and academic researchers in the quest for new knowledge and effective solutions. We discuss how biases are introduced by design in the workplace through technology and policy decision-making, how legal protections can become ambiguous through lack of definition, thus enhancing cyber-criminality, and demonstrate weakness in how the General Data Protection Regulations (GDPR) may adapt in light of new social phenomena and cultural change. We then propose legal applications with technical solutions that benefit societies in the addressing of core social operational/organisational themes and objectives, such as equality, diversity, gender pay-gap, racism, and the encouragement in the recruitment of women. This is undertaken from a combination of BD and AI ethical application perspectives, with a set of amalgamated criteria, the findings of which help identify factors for utilising in the design of a more ethically sound User Interface (UI). Three additional key socially controversial use-cases are presented (i.e., cyberstalking, sharenting, and, recruitment bias) and assessed alongside the set criteria to highlight current techno-social challenges, metrics, and applications that augment swift action in the mitigation of individual user risk. Thus, an ethically based UI Data Pipeline (VDaaS) is proposed in ensuring ethical, legal, technical objectives, and operations are met from a user perspective.

Keywords Ethics · AI · Big Data · GDPR · Bias · Privacy

V. Wylde (✉) · E. Prakash · C. Hewage · J. Platts
Cardiff School of Technologies, Cardiff Metropolitan University, Cardiff CF5 2YB, UK
e-mail: vwylde@cardiffmet.ac.uk

E. Prakash
e-mail: eprakash@cardiffmet.ac.uk

C. Hewage
e-mail: chewage@cardiffmet.ac.uk

J. Platts
e-mail: jplatts@cardiffmet.ac.uk

1   Introduction

With the promise of solutions to legal, technological, and ethical challenges, it has long been recognised that Information Communication Technologies (ICTs) also have significant impacts upon the social and economic aspects of societies [1], to the extent that the regulatory supervision, ethical, and social assessment are a necessity in meeting legal requirements. At this time, the observation of two of more of these ICTs add great benefits and can also have undesirable impacts upon human rights and ethics, further sped-up by novel ways of collecting and processing via Big Data (BD), Artificial Intelligence (AI), into "Smart Information Systems" (SIS) [2].

In utilising the self-driving car for instance, the sensory technology and AI system can cause harm when deemed unreliable [3] when a pedestrian was killed whilst using the road because of an algorithmic failure to respond to the sensor (i.e., not detecting pedestrian). Examples of SIS include Google Translate and Search engine, recommendation systems [4] and Alexa from Amazon, GPS enabled Smartphones, likes from Facebook, Policing systems (predictive), healthcare and surgery robotics, automated data sharing, augmented and virtual reality, personal fitness apps, to name but a few. All of which have wide-ranging data analysis implications concerning social media traffic into advertising and energy conservation via predictive analytics [2]. This includes questions around the traceability of recent companies and organisational sustainability reports and the measuring of own sustainability goals [5].

In the modern era, BD and analytics present opportunities that present both good and bad outcomes. For example, in [6] highlights a US article that demonstrates medical privacy and solicitation from Kaiser Premente, United Healthcare, and other insurers in signing up the deceased for Medicare advantage plans. This shows a negative side to personal data being utilised through BD to share, buy, and sell data between public and private entities. This $92 million venture was enabled through Northrop Grumman (Defense giant), Medicare, and Medicaid Services in ascertaining future beneficiary medical disorders in concert with healthcare needs and interactions with doctors, pharmacies and hospitals, all inclusive of attaining data through social media with the aim in reducing online fraud.

Just as with news organisations in attaining reader consumption data across different platforms, this evolution requires the building of customer relationships, with personalised experiences informed by consumer data profiling [7]. However, with the recent acceleration, development, and adoption of AI, much debate around ethics are prevalent due to the increased internet usage undertaken through the Covid-19 pandemic, additional crimes such as cyberstalking and personal data privacy breaches, continue to escalate in response to homeworking and in contradiction within family responsibilities with regard to the sharing of family narratives (i.e., pictures and social media contents) in the personal, private, and public domains.

Whether in academia or industry, AI solutions and trust are required to take into account numerous dimensions to include environmental, public opinion, ethical, social and legal aspects. A multitude of guidelines and principles with toolkits are published and made available globally, however uptake and implementation has been limited amongst Small-to-Medium Enterprises (SMEs), due to a shortage of skills, knowledge, and resources [8].

In addition, in highlighting the EU—General Data Protection Regulations 2016/679 (GDPR) for example, consumers find themselves in a vulnerable position as a consequence of information asymmetries between themselves and data-driven companies, therefore in mitigating these asymmetries, GDPR in practice is unable to effectively deliver full protection and transparency due to the lack of identification in the unique characteristics of dynamic data-driven business models [9]. Inevitably, consumers that are asked about the Internet of Things (IoT) cite privacy and security concerns, however because of the technology and benefits, consumers will not disconnect and find themselves in a privacy paradox [10].

- **Objective of this paper**: This contribution is to aid and influence public sector (i.e., government policymakers, NHS, and subsidiaries) and private sector (i.e., Businesses and enterprises [SME's]) in an era of Covid-19, for future data governance and practices in informing data-driven decision making with regard to key social media ecosystem commonalities (i.e., hate-speech, discrimination/bias, and to enhance public health outcomes, demonstrated with an effective values driven end user interface (UI) data pipeline).

  The research undertaken builds upon the previous works of [11–15] and aims at identifying common principles in assisting and making BD and AI systems more trustworthy in technology domains, and to ultimately help researchers grasp topics for future research.

  The paper presented also aims to assist technology development teams in giving brief and clear information derived from normative ethics and applied ethics applications.
- **Problem Description**: As part of the utility and development of VDaaS (Fig. 1), the user interface requires the design and implementation of an ethical system of operation, to help in navigating and to drill-down to identify what is utilised in augmenting risk-mitigation concerns and to ultimately ensure the effective, safe, and secure use of VDaaS. In [16], the authors highlight the widespread deployment of digitalisation across societies and the emerging new ethical challenges as a consequence of technological development.

  Therefore, managerial decision-makers, designers, and software developers are being expected to involve and to consider normative evaluations and applied ethics when producing/building new digital products. However, from within academic literature it is difficult to find ethical guidance without coming into contact with multi-branches of applied Ethics.

The chapter structure is as follows. Section 2: Looks at the application of ethics to a given technological domain from a normative and applied ethical point of view. This includes a three-tiered framework to enable ethical-technological guidance for tangible and intangible resources in decision-making optimisation and development in the domains of BD and AI. Section 3: Brings attention to the challenges that compound

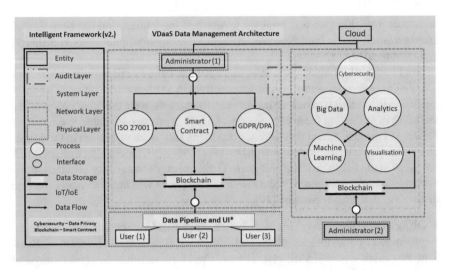

Fig. 1 VDaaS: data pipeline and UI [11]

practices such as cyberstalking, deployment of algorithmic bias, and the relatively new phenomena, "Sharenting". Three BD and AI amalgamated criteria are identified and utilised in the assessment of these challenges for solutions.

Section 4: A main theme is the continued debate of legal definitions and granularity surrounding cyberstalking, legal requirements in ensuring data privacy, which also includes actions with possible solutions of technology to the application of ethical principles in workforce recruitment, and the responsibilities of parents in sharing family narratives online. Section 5: Three main challenges and solutions are identified and addressed from the technological, legal, and ethical domain/framework aspects in relation to set assessment criteria. Highlighting government mandates for Covid-19 for example, that significantly increased internet use and prevalence's of data breach, bias, and cybercrime. Here, the challenges are grouped together to better inform a UI Data Pipeline design and implementation framework.

Section 6: Demonstrates preventative measures and protections with vulnerable groups in taking on-board design perspectives in UI data pipeline. This also highlights how societies can be negligent in their duty of care over children in sharing online family narratives, and how legal definitions and ambiguity do not help in the seeking of justice from crime, nor the application of fairness principles in organisational recruitment and decision making objectives.

Section 7: Here, particular key points are raised for future research directions and application. A main theme is the lack of robust objectives, legal definition, and clarity when addressing issues such as cyberstalking, sharenting, and human resource recruitment strategy.

2   Previous Work

2.1   Normative Ethics

As a branch of philosophy (moral), normative ethics broadly looks at what is right and wrong; or how we should act by situations presented daily such as traffic rules and politeness, with the expectancy of other people and ourselves to abide by a common set of implicit rules [17, 18]. From these behavioural contexts (i.e., expectation), reasonable actions and decisions are formulated [18, 19]. Therefore, values and shared goals are in alignment, hence our behaviour is not something that often needs any additional thought [20, 21].

In regard to everyday practices, own demands and those of others do not clash as attitudes (i.e., virtues) develop over time to enable mutually desired and complex behaviours, the questioning of which (i.e., normative demand adequacy) is necessary, and forms the core of critical and moral thinking [21]. A main purpose of normative ethics is to determine how basic moral standards are justified and formulated, of which answers can be categorised by two categories: consequentialist, or deontological and teleological [22].

2.2   Applied Ethics

However, applied ethics is usually refers to the philosophical ethics that are utilised in the moral and practical application of principles and considerations, with a focus set on real-world scenarios and challenges, and actions that are applied in a concrete setting (i.e., good or right). In the application of applied ethics is usually the practical arm dealing with normative ethics application [16] alongside considerations and treatments of moral practices, problems, professions, personal life policies, government and technology.

Conversely, regarding pure traditional ethical theory-concerned with issues such as the general right and wrong criterion, applied ethics takes a more practical point of departure with journals and overviews adding nine branches, with six to eight references per branch, which overlap in different areas, and represent only a small selection of disciplines from within applied ethics [23].

The field of medical ethics for example, deals with all challenges in the domain of medial care with principles and values in a medical setting, a main principle of which is the value of healing. Every other value is subordinated. However, in the utility and application of digital devices, there are many aspects of life-world and life practices that cannot effectively be singularly identified (i.e., a single value that dominates all), in the provision of a normative orientation [16].

Therefore, in combination of the ethics, digital artifact, and domain, we can identify desired normative aspects in formulating criteria, however with the ethics of engineering and technology, it must be delimited via its subject matter and application

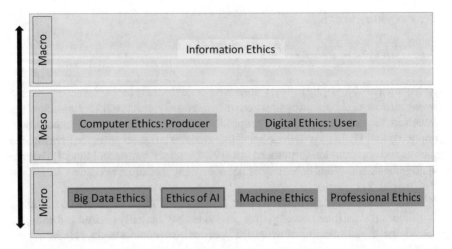

Fig. 2 Ethical domain

in a social subsystem, thus it is important regarding information technology to systematise in order to identify the topics of discussion and application in the domain of ethics [16, 24].

In [16], the authors propose a systematic three-tiered approach to help identify relevant technological aspects and domains (macro, meso, and micro). Regarding micro ethics, domain specific branches of applied ethics are utilised specifically to a certain technology or domain, hence at a smaller scale. The meso level concerns aspects that are all accounted for at a micro level, then are reflected into the context of user or producer (Fig. 2).

For example, when considering the issues of BD, and all the techno-generic (i.e., repetitive "black-box" tasks, programs, and applications) and structural values, the is also a need to focus on the effects to the user/producer. Additionally, this applies to the higher-meso level too. Here is where fundamental ontological issues can be identified.

2.3   Big Data Analytics

In [25] the author discusses organisational knowledge creation, and highlights how organisations face changes in their decision-making due to the sheer veracity, volume, and possibilities of BD. In developing the capabilities to benefit from BD, organisations and firms need intangible and tangible resources in the form of technical and managerial skills, culture, and Human Resources (HR) [26–28].

Additionally, in [29], the author highlights how organisations have optimised BD and analysis from multiple sources to aide in decision making with organisational objectives and goals. Albeit from a strategic perspective, the analysis of such data and its varieties and availability, create major benefits and challenges for firms and organisations. However, decision-making quality concepts refer to accuracy and correctness, whilst also being evaluated in an effective and efficient process [30–32].

For example, according to [33], effectiveness looks at the reliability, precision, and accuracy of results, however efficiency focuses on cost, time, and other resource aspects. Taking into consideration that since public health events such as Covid-19, consumer demand diversification has meant that higher requirements are expected from firms in terms of decision-making, and the speed at which data is accessed and processed could potentially (deliberately or inadvertently) breach an individuals privacy, permit data misuse, and compromise the sharing and use of data to include the raising of ethical issues in BD management [29].

Although, [34] states that with the ease and scale of conducting BD analytics, means that current legal-ethical frameworks are likely to completely change, therefore suggests that all stakeholders (i.e., database administrators, data engineers, and data scientists) that handle BD, should be involved is discussions of ethical BD usage and take the following five principles into consideration:

- Not to Institutionalise Unfair Biases:
  This includes racism and sexism. Machine Learning algorithms can transfer unconscious biases from a population, then exaggerate through training samples.
- Not to Obstruct User Agency:
  During moderation, analytics can determine identities and decisions. Companies need to differentiate and consider what inferences and predictions are permitted.
- Customer Transparency:
  Give a holistic view of how data is being utilised and sold, with private information controls regarding third-parties.
- Private Data Sharing Confidentiality:
  Sensitive, medical, location, and financial data shared via third-parties require restrictions on how and whether data utility can be escalated further.
- Private Customer and Identity Data:
  Private data may require legally based audits, however personal data attained with consent must not be exposed to other individuals or businesses that link back to identifying the customer.

2.4   Artificial Intelligence (AI)

In [35], the authors highlight that companies are deploying AI for personnel recruitment and selection processes increasingly, with a view to speeding-up and creating efficiencies thus streamlining the overall process. Besides AI applications and benefits, there can be direct and indirect harms to users and societies, therefore it is necessary to ensure system safety, reliability, transparency (i.e., trust). The authors in [3] review AI trustworthiness in algorithmic decision making requirements, guidelines, and frameworks

[36–39], to find a common principle set in assisting in making AI trustworthy, and to highlight the profound effect on daily lives via applications in justice, education, government, business, and healthcare sectors.

From these requirements, strategies are presented that aid in mitigating AI risks to amplify acceptance and trust in systems, with a holistic view regarding recent advances in trustworthy AI to help researchers grasp these topics and future directions of research. To minimise review inconsistencies, researchers [40, 41] analysed and reviewed the aforementioned documents and found emerging consensus on the following five additional main principles:

- Privacy
- Responsibility/Accountability
- Societal and Environmental Well-being/non-Maleficence
- Justice and Fairness
- Explainability/Fairness.

2.5 Summary

As with the all-encompassing and wide-ranging ramifications and impacts of technological development and digitalisation across societies, the ethical challenges and consequences require that managerial decision-makers, software developers, and designers, are to take on-board normative evaluations and applied ethics. As direct and indirect harm because of these technologies indicate, these issues can be evaluated in the behavioural contexts for reasonable actions and decisions to augment real-world applications, challenges, and scenarios.

In the pursuit of knowledge creation and in meeting organisational objectives and goals, the speed, access, and processing of data, can often breach an individual's privacy to include the misuse and compromising of shared data, distributed instantly across international and global boundaries. With the scale and ease of utilising big data analytics and the sheer volume of data varieties available, current legal and ethical frameworks (i.e., GDPR—Fairness and Justice) could be contributing to additional challenges to users and are likely to continually evolve at a different pace to their respective technologies and societal norms.

These challenges require that in designing any UI, the ethical system of operation should be of paramount importance, and with the combination of ethics, domain, and digital artifact knowledge, the identification of normative and applied aspects can aid in formulating a set of criteria.

We then take into account the identified key principles to demonstrate factors that affect particular behavioural contexts, and further identify technological and legal challenges, to help ascertain a novel framework of ethical operation.

From the two technological domains identified (BD and AI), three principles are derived from each domain and amalgamated to bring focus (Fig. 3) to salient use-case points as an assessment tool as follows:

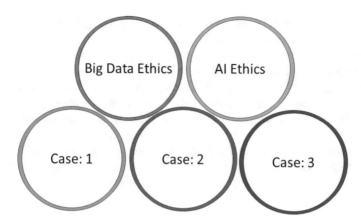

Fig. 3 Technological application domain

- Case 1: Private Customer and Identity Data: Privacy
- Case 2: Not to Institutionalise Unfair Biases: Justice and Fairness
- Case 3: Not to Obstruct User Agency: Responsibility.

3 Research Challenges and Open Problems

3.1 Case 1: Cyberstalking

In [42], the authors investigate conceptual development of cyberstalking and the right to privacy in relation to cyberbullying. In 2018, a report from the National Crime Reports Bureau (NCRB) highlighted that every fifty-five minutes there is a case of cyberstalking reported in India. For the first time in 2017, the NCRB included cyberbullying/cyberstalking data that found a total of 543 cases against women and children were recorded that year.

As with increased internet usage over time, this crime has been amplified, as a study stated that the offence in itself (cyberstalking or e-stalking) increased substantially during the lockdown period of the Covid-19 pandemic in 2020, when people where compelled to stay indoors in cyber-world [43] and through this development, gave offenders clandestine opportunities. This was reflected in the NCRB report in 2020 that found a significant increase of cyberstalking and cyberbullying was shown in recorded cases against women and children was up to 872 [42].

This form of passively monitoring causes much debate in regard to whether this is true "cyberstalking" [44], however there are also more invasive behaviours that can violate and directly intrude upon privacy, to include unauthorised social media account access (logging in) by a partner to assess activity, and to further access password protected emails [44, 45] to fabricate social media profiles and bypass privacy controls (i.e., being blocked) altogether [46].

Common targets of cyberstalking are intimate partners, yet the negative impacts of this behaviour are under-explored [47], and can be classified as a behavioural pattern of surveillance and monitoring of current or past partners [48]. In the case of the later-developed technology based Intimate Partner Surveillance (IPS) and Intimate Partner Violence (IPV), survivors of the practice have reported that the abusers utilise spyware apps, compromising of user accounts, shared cellular plans, and GPS trackers, to name but a few, in monitoring their digital activities and locations in the physical world [49–53].

Previous works also indicate that a plethora of IPS apps are widely available online [54] and being utilised against victims [55–57], of which IPV in the US for example, affects around one-third of all women, one-sixth of which are men [58].

3.2 Case 2: Racism and Bias

In the recruitment selection process of a company's workforce, occurrences of bias can affect diversity of applicants. However, a reduction in bias can also lead to a diverse company workforce [35], thus a job-role created through a bias-neutral AI will result in a pool of diverse applicants [59]. This means that with a data-driven assessment can lead to the hiring of nontraditional applicants (i.e., strong skills/from a non-elite university), which AI-augmented tools may provide better job market access for people from a wide-range of diverse socioeconomic backgrounds. In addition, in utilising AI-augmented recruitment can also introduce an array of algorithmic bias [60, 61].

For example in the context of fairness, a challenge is presented in using hiring algorithms as a recruitment system, [62] highlights that major tech company were found to be utilising algorithms that were biased against women, also in [63] the author refers to another major tech company that demonstrated bias towards black women, to include in the context of justice, the COMPAS algorithm which was found to be biased against black people as a whole, and was utilised nationally in the prediction of criminal recidivism risk.

In some cases, a training dataset may be inherently biased to a particular group based upon past outcomes being utilised to predict future outcomes. For example, in a future assessment of employees (target of interest for opportunities) the metric used is based upon number of hours worked in an office. The outcome of which could lead to the past hiring of women that work fewer hours in comparison to men (i.e., part-time work), leading to a profiling system which may indicate a less successful outcome, not be able to appropriately advertise, thus a system will imply a poorer performance of women applicants [9, 64].

3.3 Case 3: Sharenting

In [65] the authors focus on the worries of EU children compromising their privacy via disclosure of personal online data, and how paradoxically that parents share information and pictures about their children in a common practice called "sharenting". With these complexities, and in advance of sharing information, parents do not ask for a child's consent, thus favouring the the social benefits of sharenting above the potential risk to both parties.

For technologies such as social media, this has transformed the accessing and sharing of information, which also fundamentally changes the roles and responsibilities of parents and their children's digital privacy. These online behaviours (parents and children) are part of an ever-growing body of research, that includes the creation of digital narratives, personal information, and pictures, of which often are used in highly commercialised contexts [66–69].

One consistent factor that is present from within these dynamics, is the power exerted over children by their parents, which includes not just controlling their child's online behaviour, but also includes how parental behaviour on social media ultimately ensures a child's right is respected when it comes to privacy [70].

This has a positive outcome in supporting family discipline and responsibility, however, as parents frequently share their digital narratives regarding their children and families via platforms such as Instagram, Twitter, and Facebook, a permanent digital footprint record is produced. As with any new technology throughout history, the concerns for the safety, risks, and harm to children continue to be researched, the internet being no exception. These sources of potential harm include grooming, cyberbullying, identity theft, and access to adult (sometimes violent) content of a sexual nature [70].

3.4 Summary

In case 1, as demonstrated by India's NCRB, the instances of cyberstalking and bullying are said to be on the rise from 2017 until present day, primarily against women and children. This was further enabled and intensified by the Covid-19 pandemic, where global populations were mandated to stay indoors by their respective governments, which then gave rise to potential offenders with the opportunity to utilise the availability of internet apps, and services in augmenting potentially illegal digital and local surveillance efforts.

However, in the cases of victims of IPV and IPS (mainly female), debate over the granularity of cyberstalking and legal definitions continue to evade effective enforcement provisions concerning the legal right to privacy, thus denying survivors the additional ethical right to justice.

With case 2, the recruitment processes of a given institution, workforce, and company, the potential for bias is inherent in any system utilised in the pursuit of a legally equitable and ethically sound outcome. AI tools can offer effective and robust mechanisms in resource management, however the sources of recruitment (i.e., Databases) and the metrics utilised in ascertaining employee recruitment fairness, provide additional challenges concerning what is acceptable from a large company's objective perspective, and what is affordable, in terms of enabling a population with prudent access to job-markets.

As recruitment, staff retention, and relevant BD/HR management policies are enacted, recorded and utilised, from within an AI training dataset standpoint, this can directly affect the performance and credibility of a company in terms of its sustainability regarding justice and fairness to women, minorities, and those from various deprived socio-economic backgrounds.

Finally with case 3, the ever-growing amount of internet users, the phenomenon of sharenting brings attention to how children disclose their personal data, and how parents are responsible for their children's online behaviour and access with regard to sharing digital narratives (i.e., pictures of families/events) via social media platforms. Some parents can be said to be making decisions that violate a child's right to privacy (i.e., not asking for consent), thus a parent can be neglectful of their GDPR inferred legal duty of care, and unknowingly expose themselves and others to further online risks such as grooming, identity theft, and inappropriate access to adult media (i.e., sexual and violent).

Ascertaining who is ultimately responsible in their own agency when accessing the internet is also part of a growing body of behavioural research. However, when assessing the potential online risks, social benefits are stated to mainly outweigh the high-risk possibilities.

4   Research Actions and Possible Solutions

4.1   Case 1: Cyberstalking

Today, intermediate platform providers (web companies) could be said to be partaking in a form of cyberstalking in pursuit of generating revenue via a personal data choices and preferences. Even though on the face of things that the internet platform is assumed to be providing a free service, in reality users are unknowingly paying via their privacy [42]. For example, when a user browses for products and services, then products of a similar nature start to appear throughout additional e-platforms. This is because online search engines record all user inputs and distribute the data with other companies for commercial gain [71].

This practice is magnified for example in India, as the legal definition and laws surrounding cyberstalking are dissonant, as few recognise or report online stalking occurrences, and the Covid-19 pandemic has also contributed to increased episodes of physical stalking, to online virtual stalking, due to the volume of apps, spyware, stalkware, and social media tools [43]. However, in regard to future research and how people gather information on their partner, [47] suggests that a multidimensional conceptualisation of cyberstalking behaviour (i.e., passive, invasive, and duplicitous) method should be utilised as an assessment metric by trusted parties, as it is reasonable to expect the deceitful and invasive forms of cyberstalking cause more of an impact to the target.

In addition, [46] bring attention to the IPS threat model [52, 54], and offer cybersecurity experts ways in preventing their work from IPS misuse, and should consider [52] for the concepts of User Interface (UI) bound adversary for design teams to also consider.

As a recommendation, there is a need for a more holistic approach in sharing and changing account access, hence platforms may consider providing a notification mechanism for addressing impactful events (i.e., divorce, breakups, and potential breach) similar to the notification mechanisms deployed by companies when there are cases of identity theft, change of address, or even lost credit cards occur.

A series of steps could then be initiated by a tech platform to give advice in remedying the complicated entanglements to help ensure their accounts remain secure. As mentioned with the universal principle of design [72], these specific instances can improve the overall operations and utility of platform for all to include IPV victims [57].

4.2 Case 2: Racism and Bias

In an article from the Universitat Oberta de Catalunya, the question is asked: "Do men and women still use digital technologies in different ways?". Studies have indicated that with the prevalence of technology for women and men, women tend to utilise technology in different ways, which leads to the conclusion that women are not so involved in the production and design of technology. This includes those women that are older, and from disadvantaged and rural backgrounds, meaning that the same older and rural women utilise the same technology less proactively than men [73].

In response to this challenge, the United News of India, the Indian Institute of Technology Madras (IIT-M) launches a project in reducing the gender pay-gap in Wikipedia. The initiative called "Hidden Voices", launched in partnership IIT-M alumni and with the Robert Bosch Centre for Data Science and Artificial Intelligence (RBCDSAI), and Superbloom Studios (business consultancy firm), the initiative founders aim at making a positive impact on digital sources and their gender representation, thereby setting auto-generating biographies of prominent women within the coming year, notably prior to International Women's Day on the 8th of March 2023 [74].

The generated biographies include north American and Indian women that have contributed significantly in STEM fields and business domains, also the team plans to expand in areas of expertise, geography, and for application to other underrepresented communities. Due to the Natural Language Models that depend on open web datasets, and the interaction of consumer services (including Wikipedia), there are many layers of complexity to solving equal representation. In addition, as the state-of-the-art Automated Language Processing has majorly advanced, there are still instances of AI errors, especially when processing documentation around underrepresented populations [75].

4.3   Case 3: Sharenting

In acknowledgement of perceived child vulnerability, GDPR offers increased data protection in recital 38:

> children merit specific protection with regard to their personal data, as they may be less aware of the risks, consequences and safeguards concerned and their rights in relation to the processing of personal data

Also, stated to be paving the way for universal data privacy, GDPR "entrenched privacy with trust as it's cornerstone" [76] was introduced with emphasis on safeguarding teenager online activity. Alongside such measures includes the provision for consent when processing the data of minor children, prohibition of their biometric data, their right to data portability, erasure, and the right to be forgotten.

However, little attention was focused upon the identity and privacy safety of young children, whose parental postings mean that a child's private identity is slowly being eroded online. A further complication is recital 18 of GDPR that exempts personal activities and households from the protections and constraints of GDPR, a main personal online activity of which evades GDPR, in sharenting, said to be growing and widespread. Research shows that the practice of sharenting benefit parents, however it is at the cost of privacy and that of their child's private identity [77].

GDPR's inferences on parents being responsible could be counterproductive, this may augment a child's decision in not seeking parental permission, lying from a child regarding their age, thus making online activities out of the scope of parental protection. In addition, in assuming that parents are "gatekeepers" regarding their children's privacy and digital safety, fails to acknowledge that parents may not be technologically aware and/or computer literate. Additionally, there could be a danger regarding vulnerabilities referred to in recital 38 which may hinder children's ability, however the opposite is expressed as a dynamic phenomenon regarding children's rights potential [78].

These protections apply to the use of child personal data for marketing, creation of user profiles, and the collection of personal data for services directly offered to a child. However, GDPR recital 18 states that the processing of personal data by a natural person, does not apply to household and purely personal activities, hence no connection with commercial or professional activities. This means that with GDPR there is no provision for the processing of household/personal data (i.e., images) or oversight.

This means that recital 18 has questionable compatibility issues with recital 38 protections, thus differential regarding family dynamics, as images are more shareable, visible, and durable, with known and unknown audiences, meaning that it is difficult to know where parental responsibility ends and child protections begin [79].

Further research is needed to distinguish the types of sharenting alongside the potential ramifications and implications for young people and children's privacy rights [65], as a point of contention in GDPR regulation is at the lack of protection in child private identities of which pictures are shared via social networks [77].

4.4 Summary

In case 1, since the volume and availability of social media apps uptake have significantly increased alongside the Covid-19 pandemic, many platforms have benefited in the sharing of personal data across multiple third-party sites, thus boosting their revenue streams. To the user(s) this can come at a cost in terms of privacy, and increasing cyberstalking/cyberbullying figures, as these platforms record and store customer data (i.e., Big Data).

However, due to a lack of cyberstalking awareness and the legal clarity of definition coupled with ongoing debate, the reporting of such online behaviours remains low, potentially as of consequence to what constitutes "true" cyberstalking concepts. A multi-dimensional approach may shed further light upon narrowing down particular instances of cyberstalking, in addition to providing definition in mitigating harms to future victims and survivors.

Case 2 research continues to demonstrate that men and women utilise technologies differently. Examples show that women utilise technology in different ways. However, with the growing number of initiatives to attract more women into technology disciplines, the likelihood of changing potential demographic imbalances could be realised through BD and AI. The idea is that with the inherently bias database, of which is primarily made-up of men, can be auto corrected and representative. The data can then be extrapolated and apportioned more equitably in the training data of an AI, thus give more inclusive, balanced, and with more wide-ranging results to include the recruitment of women in the workforce.

Finally, Case 3 shows with "sharenting" that there is no legal definition for the phenomena at this time, however GDPR infers parental responsibility and offers increased child vulnerability protection to include the processing of data for minor children, with safeguarding teenager activities online in recital 38. A main purpose of which is the provision for the creation of profiles, collection of data, and marketing etc. that facilitates the agency of interacting with platforms and access to media (pictures, music, and text documentation).

However, in the act of a parent sharing their family events, recital 18 exempts personal home activities from GDPR protections. In the case of the Covid-19 pandemic, as with other cases, vast swathes of the population were compelled to staying at home, therefore with an increased commercialisation of data became more likely, of which GDPR protections would seem inadequate under these circumstances. Therefore, more research is needed in mitigating child identities and personal data sharing risk, as recital 38 has compatibility challenges with the intent and purpose of recital 18.

5   Discussion and Conclusion

After assessing the normative and applied aspects of ethical, legal, and technical challenges alongside an amalgamated consensus of set criteria from BD and AI domains, it was found that current applications of BD and AI technologies with use-case scenarios, that many have direct and indirect health, environmental, digital ecosystem, and social challenges.

Of these findings, many were also identified and demonstrated in terms of highlighting risks, reasonable actions, and showing glimpses of possible solutions. However, with mandatory compliance from governments for populations to stay at home due to a global public health emergency (i.e., Covid-19 pandemic), internet usage significantly increased. Therefore, the pandemic augmented themes that were already prevalent. From a human rights perspective, ethical frameworks such as justice and fairness were a primary focus and assessed at the behavioural level to ascertain these pertinent challenges.

Technological challenges were also clearly expressed in relation to influencing ongoing societal applications, and ramifications of the collection, storage, and use of personal data, via companies, organisations, and internet platforms such as social media. This brings attention to requirements of balance between protecting user from harm and free expression, when it comes to the design of policies for allowable content on social networks for example. In addition, that the subtleties of context, subjectiveness, linguistic, and cultural norms present ongoing challenges when using AI in the detection of potential violating content [80].

However, from the utilisation of BD databases and the outcomes for effective recruitment strategies, the management practices that initiate organisational/company pre-set AI algorithm training decision making tools, can too often reveal bias inherent to the technology design, which is then further enhanced by active decision making metrics at operation levels thereby potentially eroding trust, hence affecting underrepresented groups, ethnicities, and those from deprived socio-economic backgrounds.

Legal challenges were also identified with an appropriate legal framework to find current gaps in the utility of GDPR personal data privacy. Recital 38 and recital 18 for example was found to be incompatible in certain use-case scenarios, mainly blurring the line between GDPR inferred responsibilities of parent/child's consent, and between the application of private and public marketing, commercial, and data collection purposes.

Moreover, it was also clear that many of the augmented ethical principles were inexorably linked and therefore interchangeable in support of a given application of technology. A dominant theme of which was a lack of legal definition and granularity with newly evolving phenomena (i.e., Sharenting and Cyberstalking). Additionally, women and children were also found to be mainly affected by the ecosystem of which a given technological application is deployed. However, to help mitigate future risk and exposure to any application misuse, an effective UI is of critical importance as the first port-of-call to potential users, and the processing, storage, and gathering of data.

5.1 Case Challenges: UI and Data Pipeline

Challenge 1: Cyberstalking

1. Awareness and the legal clarity of definition
2. Reporting of violating behaviours
3. Unauthorised social media account access (logging in) by a partner to assess activity
4. Access password protected emails
5. Fabrication of social media profiles and bypassing privacy controls (i.e., being blocked).

Proposed Metric:

- Warning Mechanism: Viewing of Profile
- Reporting Mechanism: Abuse/life-changing Occurrences
- Record and Monitor Mechanism: Potential Passive, Invasive, and Duplicitous Behaviour [multi-dimensional approach]
- Tech Company Mechanism: Advice and Guidance to Secure Computer Account.

Challenge 2: BD and AI Bias

1. Effective Representation in AI Training Data
2. Initiatives to attract more women into technology disciplines.

Proposed Metric:

- BD Database Sourcing Initiatives
- Extrapolate and Apportion More Open-Source Data
- Auto generate Biographies of Appropriate Women
- Promote Shared-Access to Initiative Databases.

Challenge 3: Sharenting

1. Leaking Digital Narratives
2. Leaking Personal Identity Information
3. No legal definition for the phenomena
4. GDPR Infers Parental Responsibility
5. Recital 18 exempts personal home activities from GDPR protections.

Proposed Metric:

- Filtering (data cleaning)
- Public and Private Mode
- Mechanism of Notification
- Warnings in terms of freely giving personal data
- Warnings regarding acts of sharing mechanism (i.e., children's consent).

5.2   Pipeline Metrics—Measurement and Enforcement

The above-mentioned challenges present themes that will be grouped together, thus focusing on minimal steps in producing an effective UI and pipeline architecture. The focus of the discussion will therefore be "Sharenting" and "Cyberstalking/Bullying" groupings due to the significant impacts and effects from within a digital ecosystem, its society, and at an individual risk perspective.

In ascertaining individual violations (see Fig. 4), key social media ecosystem commonalities are identified and grouped together, primarily focusing on the analysis of user past entries to utilised over a wide field of research, development, and applications adapted from [80]. The below controls and measures represent actions and protections as far as reasonably practicable and are not an all-encompassing solution.

Consequently, [3] highlights that [36] presents three additional guidelines in making AI systems more trustworthy: Robustness, Ethical, and Lawful. Requirements such as accountability, explainability, and fairness, continue to be researched to facilitate and give different approaches in making systems more trustworthy.

5.3   Responsibility and Accountability: File Access Permissions

As shown in (Fig. 4 Activities), and in Linux/Unix/Windows Operating Systems, and more recently applied in [81] for identity-based policies actions at Amazon Web Service (AWS), there are three permissions defined as read, write, and execute which serves as a powerful authority restrictive security mechanism. **Read** gives user authority in

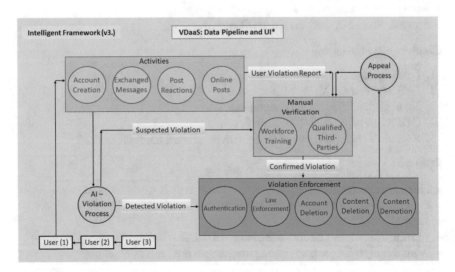

Fig. 4   VDaaS: data pipeline and user interface* [11, 80]

opening and reading files alongside directory permissions to list contents. With **write**, gives the authority to modify file contents (i.e., rename, add, and remove) on the directory. And finally, **execute** gives authority to run a program (in Windows, a .exe file). This directory and file permissions structure serves as a robust method of security and data control whether at home or utilising the public internet.

In safeguarding the right of a data subject (public and private), GDPR article 6 states that every act of data processing must have a legal basis, thus making the data controller (Fig 4 Violation Enforcement) responsible to ensure a basis and declaration to determine the legal grounds of application [82]. As article 24 states, technical and organisational measures must be ensured to enable and make certain that the controller complies with all GDPR data security and data protection provisions, and the ability to demonstrate these implementations at any given point. This means that these obligations solidify article 5(2) in establishing accountability as one of the principles of lawful processing. However, demonstrating compliance means that companies have to comply with relevant GDPR principles and accountability mechanisms [83].

For example, if a user (data subject) has write permissions for accessing the file but not on the directory, thereby the user can only read, modify and save file, but not change the name on the directory. Also, for executing a program, then the user must have execute permissions (data controller). It is possible to edit code (i.e., .exe file) however this is only permissible when read and write authority is given.

5.4 Lawful and Explainable: Integrity and Severity

This means that in terms of enforcing effective integrity techniques, such as multimodal, context analysis, advances are required. As shown in earlier examples (HR, Cyberstalking, and Sharenting), metrics are needed in assessing the efficacy of such techniques in identifying content violation. Moreover, with the low frequency of particular violations and its adversarial nature, present additional challenges in designing robust metrics.

These types of metrics include: reporting rate, appeals rate, percentage of automated deletion vs manual, AI detection times prior to an action (i.e., deleting), decision of appeals and outcomes. Of the many different forms of metrics that can be tracked, prevalence is a key metric that is taken of a sample of data, and is quantified as an overall percentage of what social network content enforcement mechanisms ultimately failed to identify.

Simply put, in the calculation of prevalence of content that is shared on a network, it is more efficient to count the reported (violating posts) instances, as distinct posts are read repeatedly over time. Also, experience prevalence may also be utilised in the refining of assessing violation severity. For example, pictures of nudity can be inferred as more serious when compared to a partially clothed instance, which to become actionable in a system may require an automated tracking metric [80].

In terms of the principle of explainability, the legal text from GDPR (Article 4(1)(4) concerning the general principles of transparency, accuracy, fairness, lawfulness, and accountability (Article 5) are relevant to enhance trustworthiness in AI [84], in addition to Article 22 which gives rights to not be subject to decisions based solely upon automated decisions that produce legal or similarly significant effects, with three exceptions [85] (Fig 4 Manual Verification).

For example, conditions need to be met such as the consent of a data subject, the necessity for the performance of a contract, or an EU or national law that regulates specific cases of automated decision-making. In the event of an automated decision, the data subject has the right to contest [86], express a personal view (i.e., human intervention in the automated decision-making) [87]. Also, recital 71 gives clarity on the level of risk (i.e., type of algorithm application) to include the auditing of the algorithm and an explanation regarding the decision as a data subject safeguard (Fig 4 Appeal Process).

5.5   Enforcement and Fairness: Identification and Mitigation

Textual (semantics) understanding of post violation can also play a key role via supervised training advances (Fig 4 AI-Violation Process) as shown in [88–90]. With social media posts for example, supervised training can undertake tasks such as identifying and correcting spelling, and grasping important context tasks such as colloquial expressions and orthographic variation. In tackling the difficult task of identifying hate speech for example, [90] has become a standard architecture in understanding text, and with additional updates from [91–93], have refined this challenge amongst others in the advancement of the state-of-the-art supervised training.

Whereas other works have considered the conveyance of content [94], to include the detection of anger, arousal, and if the author has any intentions to mislead or potentially harm a a reader or particular audience. Additional works involve identifying different style analysis in the detection of fake news as shown in [95, 96]. With regard to explainability, the GDPR principle of fairness is important to ensure that vulnerable people (i.e., women and children) can benefit similarly from rights and protections from article 5.1 (a) which is a broad area, and in most cases can be context-dependant, highly politicised, and purely a subjective concept, as the global interpretation and consensus of what is deemed fair with regard to algorithmic decision-making is unlikely to transpire [97, 98].

In addition, the principle of fairness is not defined in GDPR, as this important principle shows an escalating imbalance between a data subject and data controller [99], however the article 29 Working Party (Art. 29WP), scholars, and the European Data Protection Board (EDPB) have made remedial efforts by linking this principle to awareness [100].

Finally, within the EU Agency for Fundamental Rights, GDPR fairness mandates that it is required to be processed in an ethical manner, which could be said to circumvent transparency and the need of informing a data subject [101]. In this case, the European Data Protection Supervisor (EDPS) indicated at the time that there is a need for reflection data protection and ethics, to include how the fairness principle is perceived [102]

6 Conclusion

We have highlighted a collection of real-world scenarios to identify and evaluate normative and applied ethics to ultimately demonstrate preventative measures in relation to data privacy, algorithm biases, and legal uncertainty and ambiguity. It was found that the focus on vulnerable groups (i.e., women and children) was of main importance when designing an effective UI and data pipeline in protecting users.

Sharenting, although a relatively new phenomena, demonstrates how societies can take risks in sharing personal identity data amongst friends and family, however when sharing publicly online, these risks increase significantly and contribute to precious information potentially being used commercially without protections from GDPR due to the compounding factors of the recent Covid-19 public health emergency.

Additionally, levels of legal granularity and definition are also potentially causing inefficient data privacy protections and user access to justice, as cases of sharing and data privacy breaches increased against women and children which was further aggravated by the Covid-19 pandemic which encouraged additional rises in cases via internet apps and services. There is ongoing debate around definitions of law (i.e., Sharenting/Parental Responsibility). Who is ultimately responsible?

Of those affected by this increase, cyberstalking rates also grew from delays in clearer interpretations and debate around definitions of law (i.e., Cyberstalking/Bullying), hence cases of IPV and IPS were shown to be increasing against women and children which was also further privacy rights were compounded by the Covid-19 pandemic.

However, from within a HR recruitment perspective, casual factors that enhance bias in the selection process of applicants from deprived socio-economic backgrounds were identified. As of consequence, women were included to being less involved in the designing and implementation of technological solutions. Therefore, the need for more tangible and intangible resources, such as culture and effective managerial skills were demonstrated to help in the influence of policy makers with deriving high quality value from databases that have initiatives aimed at equality being at the centre of business objectives.

These initiatives will ultimately provide more diverse and prudent access to job-markets, which can play a part in addressing issues such as the gender pay-gap, and ethnic group under-representation. In response, VDaaS Data Pipeline and UI supports these findings and has adapted and created specifications as stated. This was a primary function to ascertain and mitigate identified challenge risks into an ethical and robust framework of operations.

7 Recommendations

In future research and discussion, there needs to be a more general focus on vulnerable groups (i.e., women and children) to include clearer interpretations and swift debate around definitions of law (i.e., Cyberstalking/Bullying), hence rising cases of IPV and

IPS, will mean that justice for victims can be realised sooner. This could mean applying additional metrics and mechanisms to user policies to help isolate future potential behavioural threats. Tech companies should be encouraged to take a more proactive approach in the provision of reporting and guidance from affected individuals in utilising their services, this could be implemented when partnerships and relationships break down, and agreed in good faith upon opening a new user account.

Also, the legal definitions surrounding sharenting will need to be quantified and more formally established. As the sharing of family narratives continue due to the social benefits of which social media was designed for, this action should minimise the chances of child identity data being unknowingly utilised in a commercial setting from home. A mechanism in the user settings could include a public and private mode of operation to warn parents of the dangers in sharing media in the public domain.

Finally, the needs to be emphasis regarding database management systems and the redressing of bias and dis-proportionality. Future initiatives will encourage trust in AI and BD systems, and help demonstrate active efforts to a wider audience in addressing issues such as the gender pay-gap. Additional benefits could include recruitment being more cost-effective, due to building a social-competitive reputation from initiative such as highlighting how to address data privacy, algorithm biases, and legal uncertainty and ambiguity in the recruitment of applicants from socio-economic backgrounds. AI training data will benefit from a database that has equality principles as core values. This should affect the evolution of culture and effective managerial skills when equality is at the centre of business objectives.

Conflicts of Interest
Authors declare that they have no conflicts of interest.

References

1. Rizou S, Alexandropoulou-Egyptiadou E, Ishibashi Y, Psannis KE (2022) Preserving minors' data protection in IoT-based smart homes according to GDPR considering cross-border issues. J Commun 17(3)
2. Stahl BC, Wright D (2018) Ethics and privacy in AI and big data: implementing responsible research and innovation. IEEE Secur Priv 16(3):26–33
3. Kaur D, Uslu S, Rittichier KJ, Durresi A (2022) Trustworthy artificial intelligence: a review. ACM Comput Surv (CSUR) 55(2):1–38
4. Wilson HJ, Daugherty PR (2018) Collaborative intelligence: humans and AI are joining forces. Harv Bus Rev 96(4):114–123
5. Wanner J, Janiesch C (2019) Big data analytics in sustainability reports: an analysis based on the perceived credibility of corporate published information. Bus Res 12(1):143–173
6. O'Leary DE (2016) Ethics for big data and analytics. IEEE Intell Syst 31(4):81–84
7. Evens T, Van Damme K (2016) Consumers' willingness to share personal data: implications for newspapers' business models. Int J Media Manag 18(1):25–41
8. Crockett KA, Gerber L, Latham A, Colyer E (2021) Building trustworthy AI solutions: a case for practical solutions for small businesses. IEEE Trans Artif Intell 1–1
9. van de Waerdt PJ (2020) Information asymmetries: recognizing the limits of the GDPR on the data-driven market. Comput Law Secur Rev 38:105436

10. Johnson SD, Blythe JM, Manning M, Wong GTW (2020) The impact of IoT security labelling on consumer product choice and willingness to pay. PLoS ONE 15(1):e0227800
11. Wylde V, Rawindaran N, Lawrence J, Balasubramanian R, Prakash E, Jayal A, Khan I, Hewage C, Platts J (2022) Cybersecurity, data privacy and blockchain: a review. SN Comput Sci 3(2):1–12
12. Wylde V, Prakash E, Hewage C, Platts J (2022) Covid-19 era: trust, privacy and security. In: Privacy, security and forensics in the Internet of Things (IoT). Springer, pp 31–49
13. Wylde V, Prakash E, Hewage C, Platts J (2021) Covid-19 crisis: is our personal data likely to be breached? In: AMI 2021—the 5th advances in management and innovation conference. Cardiff Metropolitan University
14. Wylde V, Prakash E, Hewage C, Platts J (2020) Data cleaning: challenges and novel solutions. In: AMI—the 4th advances in management and innovation conference. Cardiff Metropolitan University
15. Wylde V, Prakash E, Hewage C, Platts J (2020) Data cleaning: challenges and novel solutions for big data analytics and visualisation. In: 3MT RITA—the 8th international conference on robot intelligence technology and applications. Cardiff Metropolitan University
16. Zuber N, Kacianka S, Gogoll J (2022) Big data ethics, machine ethics or information ethics? Navigating the maze of applied ethics in IT. arXiv:2203.13494
17. Strawson PF (2008) Freedom and resentment and other essays. Routledge
18. Nida-Rümelin J (2019) Structural rationality and other essays on practical reason, vol 52. Springer
19. MacIntyre A (2013) After virtue. A&C Black
20. Mittelstadt B (2019) Principles alone cannot guarantee ethical AI. Nat Mach Intell 1(11):501–507
21. Mead GH, Schubert C (1934) Mind, self and society, vol 111. University of Chicago Press, Chicago
22. Normative ethics. https://www.britannica.com/topic/normative-ethics. Accessed 08 Apr 2022
23. Applied ethics. https://www.oxfordbibliographies.com/view/document/obo-9780195396577/obo-9780195396577-0006.xml. Accessed 08 Apr 2022
24. Zuber N, Kacianka S, Gogoll J, Pretschner A, Nida-Rümelin J (2021) Empowered and embedded: ethics and agile processes. arXiv:2107.07249
25. Fredriksson C (2018) Big data creating new knowledge as support in decision-making: practical examples of big data use and consequences of using big data as decision support. J Decis Syst 27(1):1–18
26. Chen H, Chiang RHL, Storey VC (2012) Business intelligence and analytics: from big data to big impact. MIS Q 1165–1188
27. Tambe P (2014) Big data investment, skills, and firm value. Manag Sci 60(6):1452–1469
28. Gupta M, George JF (2016) Toward the development of a big data analytics capability. Inf Manag 53(8):1049–1064
29. Nair SR (2020) A review on ethical concerns in big data management. Int J Big Data Manag 1(1):8–25
30. Li L, Lin J, Ouyang Y, Luo XR (2022) Evaluating the impact of big data analytics usage on the decision-making quality of organizations. Technol Forecast Soc Change 175:121355
31. Aydiner AS, Tatoglu E, Bayraktar E, Zaim S (2019) Information system capabilities and firm performance: opening the black box through decision-making performance and business-process performance. Int J Inf Manag 47:168–182
32. Ghasemaghaei M (2019) Does data analytics use improve firm decision making quality? The role of knowledge sharing and data analytics competency. Decis Support Syst 120:14–24
33. Shamim S, Zeng J, Shariq SM, Khan Z (2019) Role of big data management in enhancing big data decision-making capability and quality among chinese firms: a dynamic capabilities view. Inf Manag 56(6):103135
34. 5 Principles for big data ethics. https://medium.com/@uriarecio/5-principles-for-big-data-ethics-b5df1d105cd3. Accessed 28 March 2022

35. Hunkenschroer AL, Luetge C (2022) Ethics of AI-enabled recruiting and selection: a review and research agenda. J Bus Ethics 1–31
36. Ethics guidelines for trustworthy AI. https://digital-strategy.ec.europa.eu/en/library/ethics-guidelines-trustworthy-ai. Accessed 29 March 2022
37. Floridi L, Cowls J (2021) A unified framework of five principles for AI in society. In: Ethics, governance, and policies in artificial intelligence. Springer, pp 5–17
38. Perner P (2011) How to interpret decision trees? In: Industrial conference on data mining. Springer, pp 40–55
39. Andrew T (2017) An FDA for algorithms. Adm Law Rev 69(1):83–123
40. Jobin A, Ienca M, Vayena E (2019) The global landscape of AI ethics guidelines. Nat Mach Intell 1(9):389–399
41. Hagendorff T (2020) Publisher correction to: the ethics of AI ethics: an evaluation of guidelines. Minds Mach 30(3)
42. Banerjee P, Banerjee P (2022) Analyzing the crime of cyberstalking as a threat for privacy right in India. J Contemp Issues Law 7(8):35–46
43. Beware! Cyberstalking is on the rise during the pandemic. https://timesofindia.indiatimes.com/life-style/spotlight/beware-cyberstalking-is-on-the-rise-during-the-pandemic/articleshow/81924158.cms. Accessed 11 Apr 2022
44. Marcum CD, Higgins GE, Nicholson J (2018) Crossing boundaries online in romantic relationships: an exploratory study of the perceptions of impact on partners by cyberstalking offenders. Deviant Behav 39(6):716–731
45. Marcum CD, Higgins GE, Nicholson J (2017) I'm watching you: cyberstalking behaviors of university students in romantic relationships. Am J Crim Justice 42(2):373–388
46. Tseng E, Bellini R, McDonald N, Danos M, Greenstadt R, McCoy D, Dell N, Ristenpart T (2020) The tools and tactics used in intimate partner surveillance: an analysis of online infidelity forums. In: 29th USENIX security symposium (USENIX Security 20), pp 1893–1909
47. March E, Szymczak P, Di Rago M, Jonason PK (2022) Passive, invasive, and duplicitous: three forms of intimate partner cyberstalking. Pers Individ Differ 189:111502
48. March E, Szymczak P, Smoker M, Jonason PK (2021) Who cyberstalked their sexual and romantic partners? Sex differences, dark personality traits, and fundamental social motives. Curr Psychol 1–4
49. Woodlock D (2017) The abuse of technology in domestic violence and stalking. Viol Against Women 23(5):584–602
50. Southworth C, Finn J, Dawson S, Fraser C, Tucker S (2007) Intimate partner violence, technology, and stalking. Viol Against Women 13(8):842–856
51. Matthews T, O'Leary K, Turner A, Sleeper M, Woelfer JP, Shelton M, Manthorne C, Churchill EF, Consolvo S (2017) Stories from survivors: privacy and security practices when coping with intimate partner abuse. In: Proceedings of the 2017 CHI conference on human factors in computing systems, pp 2189–2201
52. Freed D, Palmer J, Minchala D, Levy K, Ristenpart T, Dell N (2018) "A Stalker's Paradise" how intimate partner abusers exploit technology. In: Proceedings of the 2018 CHI conference on human factors in computing systems, pp 1–13
53. Freed D, Palmer J, Minchala DE, Levy K, Ristenpart T, Dell N (2017) Digital technologies and intimate partner violence: a qualitative analysis with multiple stakeholders. Proc ACM Human-Comput Interact 1(CSCW):1–22
54. Chatterjee R, Doerfler P, Orgad H, Havron S, Palmer J, Freed D, Levy K, Dell N, McCoy D, Ristenpart T (2018) The spyware used in intimate partner violence. In: 2018 IEEE Symposium on security and privacy (SP). IEEE, pp 441–458
55. Roundy KA, Mendelberg PB, Dell N, McCoy D, Nissani D, Ristenpart T, Tamersoy A (2020) The many kinds of creepware used for interpersonal attacks. In: 2020 IEEE Symposium on security and privacy (SP). IEEE, pp 626–643
56. Havron S, Freed D, Chatterjee R, McCoy D, Dell N, Ristenpart T (2019) Clinical computer security for victims of intimate partner violence. In: 28th USENIX security symposium (USENIX Security 19), pp 105–122

57. Freed D, Havron S, Tseng E, Gallardo A, Chatterjee R, Ristenpart T, Dell N (2019) "Is my phone hacked?" Analyzing clinical computer security interventions with survivors of intimate partner violence. Proc ACM Human-Comput Interact 3(CSCW):1–24

58. Smith SG, Basile KC, Gilbert LK, Merrick MT, Patel N, Walling M, Jain A (2017) National intimate partner and sexual violence survey (NISVS): 2010–2012 state report. National Center for Injury Prevention and Control

59. Will AI remove hiring bias?: strategic HR review. https://www.shrm.org/resourcesandtools/hr-topics/talent-acquisition/pages/will-ai-remove-hiring-bias-hr-technology.aspx. Accessed 08 Apr 2022

60. All the ways hiring algorithms can introduce bias. https://hbr.org/2019/05/all-the-ways-hiring-algorithms-can-introduce-bias. Accessed 09 Apr 2022

61. Algorithmic equity in the hiring of underrepresented IT job candidates. https://www.emerald.com/insight/content/doi/10.1108/OIR-10-2018-0334/full/html. Accessed 09 Apr 2022

62. Amazon scraps secret AI recruiting tool that showed bias against women. https://www.reuters.com/article/us-amazon-com-jobs-automation-insight-idUSKCN1MK08G. Accessed 29 March 2022

63. Angwin J, Larson J, Mattu S, Kirchner L (2016) Machine bias. In: Ethics of data and analytics. Auerbach Publications, pp 254–264

64. The impact of the general data protection (GDPR) on artificial intelligence. https://www.europarl.europa.eu/RegData/etudes/STUD/2020/641530/EPRS_STU(2020)641530_EN.pdf. Accessed 29 March 2022

65. Bhroin NN, Dinh T, Thiel K, Lampert C, Staksrud E, Ólafsson K (2022) The privacy paradox by proxy: considering predictors of sharenting. Media Commun 10(1):371–383

66. Ouvrein G, Verswijvel K (2019) Sharenting: parental adoration or public humiliation? A focus group study on adolescents' experiences with sharenting against the background of their own impression management. Child Youth Serv Rev 99:319–327

67. Leaver T (2017) Intimate surveillance: normalizing parental monitoring and mediation of infants online. Soc Media+ Soc 3(2):2056305117707192

68. Fox AK, Hoy MG (2019) Smart devices, smart decisions? Implications of parents' sharenting for children's online privacy: an investigation of mothers. J Public Policy Mark 38(4):414–432

69. Abidin C (2017) #familygoals: family influencers, calibrated amateurism, and justifying young digital labor. Soc Media+ Soc 3(2):2056305117707191

70. Barnes R, Potter A (2021) Sharenting and parents' digital literacy: an agenda for future research. Commun Res Pract 7(1):6–20

71. AI and Big Data: what does it mean for social media platforms to "sell" our data? https://www.forbes.com/sites/kalevleetaru/2018/12/15/what-does-it-mean-for-social-media-platforms-to-sell-our-data/?sh=22b6a1552d6c. Accessed 11 Apr 2022

72. Iwarsson S, Ståhl A (2003) Accessibility, usability and universal design-positioning and definition of concepts describing person-environment relationships. Disabil Rehabil 25(2):57–66

73. Women and digitalization: reducing the digital gender gap. https://blogs.uoc.edu/in3/women-and-digitization-reducing-the-digital-gender-gap/. Accessed 25 March 2022

74. IIT-M launches project to reduce gender data gap in Wikipedia. https://www.uniindia.com/story/IIT-M-launches-project-to-reduce-Gender-Data-Gap-in-Wikipedia. Accessed 25 March 2022

75. IIT-M initiative to reduce gender gap in digital sources. https://www.deccanherald.com/national/iit-m-initiative-to-reduce-gender-gap-in-digital-sources-1094640.html. Accessed 25 March 2022

76. Buttarelli G (2016) The EU GDPR as a Clarion call for a new global digital gold standard

77. Donovan S (2020) 'Sharenting': the forgotten children of the GDPR. Peace Human Rights Gov 4(1)

78. Kišūnaitė A (2019) Children's rights protection in the EU: the need for a contextual perspective. Peace Human Rights Gov 3(2)

79. Blum-Ross A, Livingstone S (2017) "Sharenting," parent blogging, and the boundaries of the digital self. Pop Commun 15(2):110–125

80. Halevy A, Canton-Ferrer C, Ma H, Ozertem U, Pantel P, Saeidi M, Silvestri F, Stoyanov V (2022) Preserving integrity in online social networks. Commun ACM 65(2):92–98
81. Zhang B (2020) AWS identity-based policies with "Read", "Write" and "Execute" actions. Master's thesis, University of Waterloo
82. Kurtz C, Wittner F, Semmann M, Schulz W, Böhmann T (2022) Accountability of platform providers for unlawful personal data processing in their ecosystems-a socio-techno-legal analysis of Facebook and Apple's iOS according to GDPR. J Responsib Technol 9:100018
83. Vedder A, Naudts L (2017) Accountability for the use of algorithms in a big data environment. Int Rev Law Comput Technol 31(2):206–224
84. Hamon R, Junklewitz H, Sanchez I, Malgieri G, De Hert P (2022) Bridging the gap between AI and explainability in the GDPR: towards trustworthiness-by-design in automated decision-making. IEEE Comput Intell Mag 17(1):72–85
85. Kuner C, Bygrave L, Docksey C, Drechsler L (2020) The EU general data protection regulation: a commentary. Oxford University Press. https://global.oup.com/academic
86. Bayamlioglu E (2018) Contesting automated decisions. Eur Data Prot L Rev 4:433
87. Roig A (2017) Safeguards for the right not to be subject to a decision based solely on automated processing (Article 22 GDPR). Eur J Law Technol 8(3)
88. Yang Z, Dai Z, Yang Y, Carbonell J, Salakhutdinov RR, Le QV (2019) XLNet: generalized autoregressive pretraining for language understanding. Adv Neural Inf Process Syst 32
89. Peters ME, Neumann M, Iyyer M, Gardner M, Clark C, Lee K, Zettlemoyer L (2018) Deep contextualized word representations. arXiv: 1802.05365
90. Devlin J, Chang M, Lee K, Toutanova K (2018) BERT: pre-training of deep bidirectional transformers for language understanding. arXiv:1810.04805
91. Liu Y, Ott M, Goyal N, Du J, Joshi M, Chen D, Levy O, Lewis M, Zettlemoyer L, Stoyanov V (2019) Roberta: a robustly optimized bert pretraining approach. arXiv:1907.11692
92. Lan Z, Chen M, Goodman S, Gimpel K, Sharma R, Soricut P (2019) Albert: a lite BERT for self-supervised learning of language representations. arXiv:1909.11942
93. Zhuang L, Wayne L, Ya S, Jun Z (2021) A robustly optimized BERT pre-training approach with post-training. In: Proceedings of the 20th Chinese national conference on computational linguistics, pp 1218–1227
94. Rajamanickam S, Mishra P, Yannakoudakis H, Shutova E (2020) Joint modelling of emotion and abusive language detection. arXiv:2005.14028
95. Jeronimo CLM, Marinho LB, Campelo CEC, Veloso A, da Costa Melo AS (2019) Fake news classification based on subjective language. In: Proceedings of the 21st international conference on information integration and web-based applications and services, pp 15–24
96. Mihalcea R, Strapparava C (2009) The lie detector: explorations in the automatic recognition of deceptive language. In: Proceedings of the ACL-IJCNLP 2009 conference short papers, pp 309–312
97. Abiteboul S, Stoyanovich J (2019) Transparency, fairness, data protection, neutrality: data management challenges in the face of new regulation. J Data Inf Qual (JDIQ) 11(3):1–9
98. Piasecki S, Chen J (2022) Complying with the GDPR when vulnerable people use smart devices. Int Data Priv Law
99. Butterworth M (2018) The ICO and artificial intelligence: the role of fairness in the GDPR framework. Comput Law Secur Rev 34(2):257–268
100. Wachter S (2018) Normative challenges of identification in the Internet of Things: privacy, profiling, discrimination, and the GDPR. Comput Law Secur Rev 34(3):436–449
101. Handbook on European Data Protection Law. https://fra.europa.eu/sites/default/files/fra_uploads/fra-coe-edps-2018-handbook-data-protection_en.pdf. Accessed 10 Apr 2022
102. Opinion 4/2015: towards a new digital ethics. https://edps.europa.eu/sites/edp/files/publication/15-09-11_data_ethics_en.pdf. Accessed 10 Apr 2022

Law Enforcement and the Policing of Cyberspace

Alice Baraz and Reza Montasari

Abstract Today, people all around the globe rely on an internet connection to function in twenty-first century society. The evolution of technology along with society itself have led to having an online presence being an important aspect of life. Users of the internet can connect and interact with one another globally, with it being easily accessible and available. Cyberspace is a new crucial point for modernisms, social media and businesses, bringing efficiency and enjoyment to many. Juxtaposed to this is the opportunity this connectivity affords online criminality to take place. Cyberspace poses significant challenges for law enforcement, where generic policing strategies can not necessarily be applied. The 'non-physical' nature of cyberspace leads to issues when everyday police forces try to govern the digital world and units must steer away from their usual day-to-day operations, in this fast-developing arena. This chapter will analyse the different challenges that police, and law enforcement agencies face when attempting to police the cyber world and protect the public from the ever growing and evolving risk of cybercrime.

Keywords The dark web · Digital policing · Deep web · The Onion Router · Digital forensics · Bitcoin

A. Baraz (✉)
Serious Fraud and Cyber Unit, Hertfordshire Police HQ, Stan Borough Road, Welwyn Garden City AL8 6XF, UK
e-mail: Alice.Baraz@herts.police.uk

R. Montasari
Department of Criminology, Sociology and Social Policy, School of Social Sciences, Swansea University, Swansea, Wales, UK
e-mail: Reza.Montasari@Swansea.ac.uk
URL: http://www.swansea.ac.uk

1   Introduction

Cyberspace is a virtual world that is created by networks, routers, servers and internet enabled digital devices [1]. Gibson describes cyberspace as 'the creation of a computer network in a world full of artificially intelligent beings'. This was within his science fiction book, it is now a technically used term [2]. Cyberspace was initially seen as a place in which chat rooms instant messaging and online games originated. However, the term cyberspace has now been implemented in more negative ways, with online cybercrime and dark web emergence. Although cyberspace is still a place which is a huge part of modern social culture, blogs, discussion boards and web-based platforms such as the dark web have emerged. Cyberspace is a protective source of free speech, free search and the ability hide and disguise location and identity, which is desirable for those who wish to remain anonymous online. Certain online activity is obviously restricted, such as extreme pornography, child images and sites in which lead to sales of weapons, drugs or trade. One of the most difficult aspects of cyberspace to police is the dark web.

The Dark Web is content from the World Wide Web that exists only on dark-nets which are networks within the internet that allow potential users to access only with certain specific configurations or software. General users of the dark web are those who have private computer networks that they use in order to communicate with other users, roam, conduct business or look up images and videos. The dark web is a small section of the deep web which has been intentionally camouflaged and cannot be accessed by generic web browsers. Users of the dark web would generally use proxy servers such as The Onion Router Project or 'TOR' via an anonymous series of connections to allow users to search the dark web freely without fear of anyone knowing their identity [3]. Interestingly, the dark web was originally created by the United States Naval Research Lab to provide safe space for military units and agents could communicate without being located or identified. Strong encryption and anonymity protocols to ensure IP addresses could not be traced and that servers that run dark web sites were kept anonymous, allowed this. Aware of the large growing problem of cybercrime and both nationally and internationally, working with the private sector to combat criminality on the web [1].

TOR browser were initially designed for privacy, anonymity and security for users of the internet for both legitimate and illegitimate reasons. This is because of their layered encryption system which makes it next to impossible to track location or IP addresses of those who are using them, ultimately meaning that users are free to share whatever they desire. Because of this anonymity the dark web, alongside the technology which promotes it, is replacing the way in which crime is conducted online today. The true international crime that the dark web enables, shows cross-border evidence, perpetrators and ultimately the proceeds of crime hard to trace [4]. Technology is used by criminal organisations and individuals to go about criminal activity via the web with a masked identity. It is this convulsion that leads law enforcement to the difficulties they have in policing the dark web [5]. Alongside large popular networks such as TOR, Freenet, Riffle and I2P which are organised both individually and by public organisations, smaller peer-to-peer networks can make up the dark-nets [4].

The term 'dark-net' came about after the infamous Silk-Road scandal in association with TOR Onion Services. Silk Road was an online black market that sold anything from illegal drugs to identities bringing in millions each year. The technology such as TOR browsers was initially designed for privacy, anonymity and security for users of the internet for both legitimate and illegitimate reasons. The content of the dark web ranges in its entirety. Political forums, child pornography and Bitcoin fraud related services, with Bitcoin being the general method of payment used among the dark web. Post Silk-Road, media outlets have drawn focus to these black markets. The Silk-Road markets surfaced in 2011, the 'Diabolus' market being one of the first markets on the dark web, its contents being seized by law enforcement. The markets on the dark web have no user protection and are in danger of being shut down by law enforcement at any time [6]. However, regardless of these markets being shut down, this does not mean that others will not re-appear in their absence. There is not much to propose that the law is nearer to restricting the crime that is concealed within cyberspace. The abundance of criminals who exploit the flaws in the practices of cyberspace, in order to commit crime without being identified has led law enforcement to face major challenges. Police have used the legislation that is available to them but without the growing technology that is now available, in mind. In 2016 it was argued that there were clear legal and technological gaps that exist in the law and the ability of law enforcement to cope and respond to cybercrime due to this anonymity and encryption [4]. The crime conducted in cyberspace is leading to law enforcement both in Europe, the UK and America to alter the way that they investigate crime on the internet. Collaboration of law enforcement agencies to tackle this spectacle has been suggested however critics may argue that expanding formalities of the law without a clear indication of where it might result in an undesirable impact [7]. As a matter of fact, one in ten people in the UK are victims of cybercrime and in this day in age are more and more likely to fall victim to perpetrators of cybercrime than they to offline crime [8].

The aim of this chapter is to carry out a critical analysis of technical, non, technical, legal and ethical challenges in law enforcement and other security agencies in policing cyberspace. The chapter will explore the different methods of policing the dark web and whether the strategies used by law enforcement to combat the online crime is sufficient and successful. To this end, the following research questions will be addressed:

1. What are the non-technical challenges of policing cyberspace?
2. What are the technical challenges of policing cyberspace?
3. What are the ethical and legal challenges of policing cyberspace?
4. What are the possible avenues for future research to assist addressing the stated challenges?

The remainder of this chapter is structured as follow:

Section 2 analyses the most identifiable non-technical challenges of policing cyberspace while Sect. 3 examines the most identifiable technical challenges of policing cyberspace. Section 4 explores the legal and ethical challenges of policing cyberspace, and Sect. 5 discusses the possible future avenues that the police could adopt to successfully police online crime. Finally, the chapter is concluded in Sect. 6.

2   The Non-technical Challenges

Setting up a persona online is easy, free and consequently, desirable. It is understandable why so many people lean into the idea of existing online, whether this be to use social media, online shopping, communicating internationally or ordering a takeaway. Nevertheless, cyberspace can be a huge part of daily life for the majority of the population [9]. However, this nonchalance allows for crime to soar the internet relatively easily. Cyberspace allows for a constant threat of crime because of its open environment and the number of users that enjoy it. Cyberspace is vulnerable to an intensifying range of attacks by online criminals, terrorists and hackers from around the world. Organised crime takes the front stance in the digital world. This is because there is less risk online than there is by physically committing generic crime offline. The challenges of policing the cyber world are similar worldwide. High tech crime is real, especially in society today. It is still difficult for individuals to accept that we as a whole population should be taking cybercrime more seriously [10]. The general attitude toward cybercrime is that it is not as serious as physical crime such as robbery or burglary even though it makes it easier for criminals to commit serious crime from behind a computer screen. The distinct ignorance that the world has toward cybercrime and the capabilities that people in the 'digital society' have could possibly stem from lack of knowledge surrounding cybercrime and cyberspace, particularly the older, less technical generation [10]. The increasing extensiveness of the global reach of the internet creates a variety of fresh demands on the police and security services which challenges the customary governance over the cyber-security territory that may lead to disregarding these original customs all together [11]. The remainder of this section will explore the most identifiable non-technical issues on policing cyberspace.

2.1   The Global Reach

During the recent pandemic, it was recorded that cybercrime in 2020 increased whilst most generic serious crime rates dropped. As well as the estimated global losses in 2020 being $1 trillion and $6 trillion in 2021, being thought to cost the world annually a whopping $10.5 trillion a year by 2025 [12]. Although it is horrendous to think about, this is understandable, due to less people being on the streets due to lockdowns, company closures, ill health and fear from the general public. Criminals had to operate in another ways. Criminals were more or less forced into continuing their work via online domains rather than physically in person. The online world allows for way more opportunities for distributed networks that work peer-to-peer around the world. It is easier to communicate globally and allows for a further reach, even if only one person is stimulating the original crime. The online world permits someone from say Russia, to contact someone in the middle east who contacts someone in the UK and passes it on to someone in Europe [13]. This is how those who have interest in terrorism and terror groups communicate and pass on information and instruction to those abroad. This global reach allows for a lone person or 'single agent' to spread whatever they desire, to complete simple or complex criminal activity. Advantages of online criminal activity in comparison to the physical practise of crime on the street is that the costs of doing so are exceptionally low. Many criminals taking advantage of weak cyber security in certain countries in order to go about being the first base for crime [13].

Anyone with access to a computer, phone, laptop for example, and the knowledge of what they intend to distribute poses a threat on society. As well as this, the single most desirable feature that comes with online crime is the allowance of conducting this activity with a blanket of anonymity. In reality, the likelihood of leaks occurring of those involved is very low. Therefore, many criminals prefer to hide behind the computer screen as opposed to having physical involvement, whether it be one savvy mastermind or many perpetrators at once. Circumstantially, many everyday crimes such as theft, blackmail, murder for hire have evolved to exist online [14]. It can be argued that the sheer number of users that browse the web and exists in cyberspace is completely and utterly impractical in regard to policing each and every crime that is reported in a thorough manner [15]. There is a constant growth of users of the internet, and it will be ever growing as understanding and knowledge of using the digital world evolves and more and more every day acts exist online rather than physically. Identifying each and every spam email or phishing link would be near enough impossible.

2.2 Hybrid Cybercrimes

If asked to name some traditional crimes that happen in day-to-day life around the world, most of the population could give you a list of generic well-known examples. Traditionally, these crimes would not be conducted via the internet or through cyberspace, an entirely new global operations that have emerged in recent years. Truthfully without the internet in place, crimes like this have been evident for decades, regardless of whether these are happening physically offline or online, just maybe with a more localised or possibly national reach. It is just a matter of 'true cybercrimes' being a product of the larger opportunity that the online world provides [14]. The silver lining that stem from this is that there is already existing law enforcement experience from professionals regarding these offences that may be able to be applied. Behaviours evident online could also be linked to previous criminal justice processes and existing substantive law [11]. Hacking and cracking alongside Trojans

and viruses and denial of service (DOS) attacks are crimes which are cyber depen-
dant and would ultimately need a network to even be a crime. Although, crimes
such as spying, or vandalism have been adapted to overwhelm the integrity of access
mechanisms online.

2.3 Jurisdictional Issues

The internet is a 'giant network that interconnects innumerable smaller groups of
linked computer networks' [16]. It is a global means connecting users worldwide
with a global audience. As with any other crime, the UK have legislation that they
follow in order to prosecute a suspect for committing a crime. 'When online, one
is almost everywhere' [17]. Consequently, as cybercrime has a global reach, it is
sometimes difficult to work out jurisdictional issues that may arise. An offence in
one part of the world may not be an offence in another. Cultural differences play a
huge factor as seriousness falls differently on the scale [17]. This is the same with
criminal and civil disputes, one country may regard a crime as a criminal, another
a civil dispute. Also how do you settle jurisdictional authority, when the perpetrator
develops an attack from one country, sends it to another server which send it to a
variation of victims around the world, working out who had jurisdictional authority
would be very challenging [16]. Also, whether he who had done wrong must be
tried where they are physically found? Nevertheless, occasionally there are smaller
disputes as to jurisdictional factors. Such as in the case of *R v Arnold*, an extreme
porn case where the USA passed over to the UK to prosecute because there was more
likely a secure conviction under UK law. Although this may not be likely in many
cases [11].

2.4 Lack of Technology and Stepping Away from Usual Practice

As outsiders, many people look at the police and think, of course, they would have
the best technology and the best resources available in order to prevent and detect
crime. This is not entirely correct. The police do not have the access to technology
that would be needed in order to attack the scope in which cybercrime can reach. The
lack of technology and experience will stem from the police not having the budgetary
capabilities. Resources are not always available to police forces and responsive profi-
ciencies are not always either, for the emerging matters that existing cyberspace. Nor
can the police expedite the relevant policies or networks of security in order to do
so. Large scale investment in developing skills and technology is needed, as well as
focussing on recruiting or using cyber-aware officers in policing cybercrime [18].
The skills and knowledge now needed by police is an increasing pressure, this is

because of the capabilities that we now face when tackling crime online. After the ransomware attack by WannaCry in 2017 there was worldwide acknowledgement as to the seriousness of the attack and for growing cybercrime. This increase in cyber-crime needed to be matched with increased level of skill by the police. It is one thing to possess the aptitudes needed in order to detect and prevent cybercrime and another to utilise these capabilities [18]. Police forces are now built with a cybercrime unit which will be filled with capable officers and investigative staff working to protect the public from falling victims to attacks online. Officers and staff with appropriate specialisms will be absorbed into these units in order to form the defence towards cybercriminals and improve the response that we have as a nation. Not only that but would vastly improve both organisational and professional skills and experience across the force.

Local policing is routine and is based upon practices that have been taught and used for a long time. Having cyber specific policing stems away from the original skillsets of officers within the force. Not only can this lead to investigative issues but can consequentially lead to changes in work patterns, an imperative component of policing work. For example, a police officer knows the law and witnesses similar crime every day for his entire career, the types of behaviour witnessed may not hold similarities to those investigated in cybercrime cases, and maybe not evident in social or cultural or ethical boundaries that an officer would come across daily. 'Cyberspace is like a neighbourhood with no police department' [11]. Online crime has many differences to the daily crimes that the police force deal with, usually keeping dangerous criminals off the street is achieved with maintenance of local order, whereas new professional experience is needed in the cyber world. These misunderstandings propose less circumstances in which criminal justice policies can be created or followed, without the relevant sources of knowledge and understanding from those imposing them [11].

2.5   Under-Reporting of Cybercrimes

Cybercrime is under reported for several different reasons. A study showed that only around 120–150 out of 1 million cyber related crimes are reported to police, with most of these being less significant crimes such as minor fraud [19]. However, if a home were to be burgled or a car stolen or an assault took place, this will have been reported to the police within a short time after the incident has happened. This may stem from possible lack of trust or low expectations of the police force but also a lack of understanding from the public from what constitutes a cybercrime [20]. A general understanding, that may prove true in some situations is that the police cannot do anything regarding cybercrimes. A simple Instagram takeover, Facebook hack or email compromise will usually be dealt with on a 'protect' outcome rather than a 'pursue' outcome. Usually, officers and investigators are unable to track down a single compromise of an account and advice and guidance as to stopping it from happening again is a more likely consequence than an actual arrest or prosecution.

The potential reputational damage that a company may go through if they were to report a cybercrime, such as a virus or ransomware attack may lead to negative impacts on the company, so is preferred to be dealt with in-house rather than publicly [20]. Companies may deter from reporting because of this. Not only companies, but victims of common scamming methods such as 'phishing' emails may be outright embarrassed to admit that they have entered card details or sent money over to an account online, so also may deter from reporting as they feel ashamed [20]. Another possible reason for the under-reporting of cybercrime may be the lack of knowledge from the public as to where to report it.

Action Fraud is the main body that is used for the reporting of cybercrimes and a place for victims of fraud to go to make an official crime report. This report is then passed on to the National Fraud Intelligence Bureau (NFIB) to then be allocated to an officer or investigator within the police force related. After the crime is reported to action fraud, taken on by NFIB and passed to the police, if there is no major investigation, the police play a very small part in the process. Action fraud was born in 2009, meaning it has been around for a while, however it may be that it takes a longer time for people to become aware of the system in place as it may be that they have not before been a victim of such crime before and it could be argued that these forms of intervention are still relatively new in the policing world [11]. Now, the police have the capabilities to follow up these reports and the widespread knowledge and technology to be able to examine and investigate crimes online.

3   The Technical Challenges

Cybercrime remains the area of criminal activity that poses the most risk to society. The online world is hard to govern. There are many different methods in which criminals use the web to conduct crime. This section explores the technical challenges that law enforcement face when attempting to police harmful online behaviour and crime.

3.1   Anonymity

Realistically anyone should be able to search and ascertain information from online sources without risk of prosecution from law enforcement and government agencies. Anonymity can benefit online users in many illegal and legal particulars. Those who wish to have an online persona, that differs from who they are, just to protect their identity and surf the web freely are not breaking any laws by doing so. However, those who choose to use anonymity online to conduct crime, for example selling contraband, phishing, hacking or spreading malware such as DOS attacks online may well be [21]. The internet allows for individuals to engage in actions without revealing who they actually are. As this may be beneficial to some, it is not beneficial for the police when undergoing an investigation. Those who wish to conceal their identity to express thoughts seen as undesirable or unpopular opinions without facing backlash are not being unlawful in their actions, maybe outspoken but not unlawful. Freedom of speech is not illegal.

There an array of tactics in which users can conceal their identity online. One being using a proxy server to do so. Proxy servers are used to connect one user with a server that the user requests resources from. These servers allow the identity of the user in order to allow the user to use the web by masking the Internet Protocol (IP) address of the user and substituting it for another IP address [21]. This makes it look as though the user is at another address, sometimes in another country and makes following this trace very difficult for investigators. Cyber criminals can hide IP addresses or encrypt internet traffic in order to go undetected when conducting criminal activity online. Many sites allow for encryption of data, such as The Onion Router better known as 'TOR' also 'Freenet' and 'I2P' [21]. These sites host different websites within them that have the capability to mask the identity of the user. Only those with encrypted data will enter. Numerous difficulties come from the masking of user identity online, this is because, an investigator carrying out an investigation will do all of the relevant checks and searches within their jurisdiction to try and find an IP address that a crime was committed from, in hopes to find a home address that a warrant can then be executed. It proves difficult to show that this IP address was in fact linked with this specific home address. Usually, cyber criminals are successful in masking themselves in this way, making it near impossible for police to catch those conducting crime in this manner.

In order to catch these cyber criminals, police must determine who conducted these crimes and what device was used to do so in order to gather relevant evidence to prosecute. Attribution techniques are used to determine this. The use of anonymity features makes this very challenging. Attribution is prevented further when more technical criminals, who are computer savvy and knowledgeable on all things cyber, use botnets to conduct crime. Botnets are digital devices that are controlled by remote access tools [21]. Botnets are utilised as another entrance for criminals to distribute malware to other devices in order to gain access and consequentially gain control of these systems. The advanced technology can gain access to a user's device without their knowledge of them being controlled or even are aware that their device has been infected. Regardless of the knowledge of the victim, the botnet may be able to gain access to personal information, passwords, bank accounts of the user in order to steal or conduct other criminal activity. The police and other law enforcement agencies use 'traceback' methods to detect where a crime originated and who instigated it by tracing the acts back to the source [21]. This is usually when the cybercrime is detected by the victim or after a cybercrime has occurred. Investigators use the resources available to them to access log files such as application logs or events or entity data. The use of their resources available may detect an IP address linked to the crime. The challenges that arise from just these methods, steering away from criminals obscuring their identity for a moment, are very evident. Using these resources is very time consuming for investigators, sometimes waiting weeks for the results to be returned. Not only this, but in order to utilise these resources there must be some knowledge ability and skill in order to obtain these results, such as training or skill-based knowledge. In the end, there may not even be an identifiable source in order to obtain a warrant and prosecute, so it can sometimes be disheartening or pointless for investigators to explore this data [22]. Alongside this, as mentioned previously, the use of botnets obscures this information anyway as well as DOS attacks and multiple user crimes.

3.2 Resources and Abilities of Cyber Units

There is an argument that investigators within law enforcement may not have the necessary equipment in order to obtain evidence even if they are able to follow the flow of a crime. Numerous digital devices have exclusive software that operates to require specialty apparatuses in order to collect, preserve or identify digital evidence. Digital forensics tools are needed to conduct cyber investigations to the highest standard. There are limited availabilities for law enforcement agencies to conduct these investigations with the resources available to them. It could be argued that specialised units may be able to investigate these a little better than other units. However, these units may be able to investigate a small number of crimes, because of under- reporting and lack of knowledge from the general public in regard to cyber crime. Digital forensic teams and the success of cyber-related investigations may be determined by the resources, skills and abilities of the law enforcement units themselves [22]. There is much speculation over the abilities of cyber investigators and argument that they are 'lifelong learners' who are continuously evolving their knowledge and following training in order to keep up with every changing technology and also changing tactics by criminals. It can be highlighted that investigators may have a short lifespan in regard to the knowledge and skills that they have obtained. Both technology and the methods, MO and tactics of cyber criminals are ever changing and are hard to keep up with [21]. Consequentially, investigators that work within the public sector may be more inclined to leave and join a more beneficial employer such as the private sector, where they may be more appreciated for their knowledge and skills as well as financial escalations. Again, leaving law enforcement agencies with a lack of resources from physical staffing issues [23]. If there are less people working in the public law enforcement sector, then the less cases will actually end up being investigated all together.

3.3 Loss of Location Data and Data Retention Issues

Carrier Grade Network (CGN) address translation technologies have been implemented widely by Internet Service Providers (ISPs) [24]. CGN technology has led to major gaps in the capabilities of law enforcement in their efforts to investigate cyber-crimes. CGN technology is used by ISPs to share solo IP addresses among numerous users. Electronic Service Providers do not generally retain information regarding the service port number, however when investigating cybercrime, the precise time of the criminal activity would be required for investigations to occur, as well as the service port number. Users cannot always be distinguished between others [21]. Cyber investigators face many challenges as a result of this. Most of the time, if the timing is wrong on any applications for source data it will be disregarded, but in other cases, investigators will be presented with a huge supply of data and IP addresses to search through. This leads to many innocent customers being involved in an investigation and delays the investigation and finding a suspect responsible.

As well as encrypted data, virtual money has now become an ongoing issue for cyber investigators. Following the money has proven much more challenging for investigators because of crypto currency being used in online criminality. Cryptocurrency is a digital payment system that does not need a bank authorisation to verify transactions. Anyone around the world can send and receive payments via cryptocurrency, as long as they have an internet connection [25]. This payment is stored in a personal digital wallet. The transaction via cryptocurrency is verified via encryption, which is why it is so difficult to track. They are also desirable to potential users due to the possibility of racking up interest from the accounts, you can also buy and sell crypto currency to would-be investors. Criminals may be convinced by cryptocurrency due to not having to provide identification when signing up for all cryptocurrency accounts. Unlike what would not have to be done in a bank, making it effectively anonymous. Anyone may own a crypto account without going through a bank or government as a third party. Everyone may access a 'blockchain' to view any transaction that is stored in a permanent and public ledger [21]. Regardless of this, the identity of the user is not always available. Some success has come from tracking cyber criminals via crypto currency but tracking a single transaction within the whole network is discouraging [25]. The time and efforts of law enforcement sometimes means that the money has already been moved out of the account before it can be chased. Hundreds or even sometimes thousands of transactions could have been made to and from single account so it can be hard to keep up. A sort of 'cat-and-mouse' game with law enforcement and perpetrator. Because perpetrators are aware that they can be traced by law enforcement, money will be moved back and forth, in and out of different wallets, in order to lose the trail [23].

3.4 The Dark Web

Cryptocurrency is the main form of payment used on the dark web. Believe it or not, it is perfectly legal to access the dark web as well as encrypted sites such as Tor. It just means that this may be used in evidence if you are part of an investigation. The layered encryption system of the dark web makes all users location and identity anonymous and untraceable. Darknet encryption technology is used to guarantee anonymity by directing all user's data through transitional servers [23]. The dark web can be used for both bad and good. The anonymity of the dark web and the secure channels that run through it are great for concealing governmental activity communications. On the evil side of the dark web is a fully functional hub for criminals. The anonymity of the dark web alongside cryptocurrency allows for, as touched on above, drugs and weapons trade, child pornography and even acts of crime for hire [23]. Training law enforcement agents in understanding the dark web and relevant evidence that may come from it has proven as a challenge for law enforcement. The lack of knowledge of the dark web by officers and lack of understanding about how criminals use it to commit crime in their jurisdictions. A workshop organised by RAND showed that there were many challenges that police faced when trying to police the dark web [23]. A few of these concerns include the followings.

Legal needs such as jurisdictional issues, as mentioned previously and also entrapment: Law enforcement authorities needing to pose as criminals online in order to obtain information surrounding online businesses could create the risk of entrapping these criminals. Ethical issues can arise from this, and officers will need to prove that beyond reasonable doubt that the defendant was not entrapped as a defence when prosecuting. Identifying crime and developing law enforcements awareness regarding the types of crime that occurs on dark web sites. Suspect identifications and officers also needing to be able to recognise articles online such as log in data in order to prosecute these criminals about links to the dark web. The relevant technical data that will be used in evidence in court in from of lay people as the jury, that need to understand this evidence in plain language: It was stated 'to that end, a high-priority need, identified during the workshop is encouraging establishment of standards for new processes used to capture dark web evidence.

3.5 Data Protection Legislation—The National Legal Framework

Many issues arise from the police needing to obtain data in order to assist in ongoing investigations. Particularly in cases where they may need to obtain data from private contingents. Legal paperwork and authorisation by a judge may be needed in order to obtain this information. This takes a lot of time and may interrupt the agenda of the police, as well as lengthening the investigations. The Data Protection Act (DPA) 2018 is the UKs implementation of the General Data Protection Regulation (GDPR), which controls how personal data is recorded and used by organisations such as service providers [26]. It also provides users with rights that they have, such as stopping or restricting the processing of data, access personal data and have data erased [26]. Governance of public and private sector data generation, collection, storage, analysis and use varies from one country to another. The United Nations High commissioner for cybercrime stated 'the increased interlinking of public and private data processing and the track record to data implying mass, recurrent misuse of personal data information by some business enterprises confirms that legal measures are necessary for achieving an adequate level of privacy protection [23]. Although this helps customers worldwide, it is not as helpful to investigators trying to obtain personal information and data as part of an investigation. Some countries have devised strong data protection laws and others have weak or no laws on data protection. Contrary to many Countries such as Ghana allow government officials to access personal data without any legal documentation [23]. However, national frameworks and the global reach of cybercrime and the internet means that is the transnational regulations that a required to monitor the free flow of personal data across borders. For this information to be released to law enforcement, a GDPR request must be given to the company that is relevant to the investigation and legally request this information on behalf of the police.

4   The Legal and Ethical Challenges

The maintenance and enforcement of law online in cyberspace poses may challenges for law enforcement agencies, mainly because crime takes place within global context that are nationally defined [27]. Doing right and doing wrong is the basis of all ethical arguments. Everyone faces ethical challenges daily, it is a part of every-day life. Hackers, alongside law enforcement, face ethical frameworks and the nature of 'computer deviance' is equivocal and intricate and is usually the culprit, although it may be obvious to the naked eye [28]. It is important that security, personal privacy and technology are the main factors that ethical issues arise from and it is important that we know and understand why and how. This section discusses the ethical arguments behind policing cyberspace and suggests where the line could be drawn in right and wrong approaches with a particular focus on the dark web. Furthermore, this section examines the legal frameworks that are available and the arguments that stem from the current legal structure.

4.1   Police and Ethical Challenges

The dictionary definition of ethics is 'moral principles that govern a person's behaviour or the conducting of an activity'. Ethics is the argument between doing right or wrong in a moral sense. Choosing between two options that may both lead to undesirable outcomes. Codes of morals or standards of conduct in a career or a profession are examples of ethical factors. Although what is unethical is not necessarily unlawful. In law enforcement there is a code of ethics that must be followed in all decision-making aspects of the job. Figure 1 is a redrawn visual representation of the national decision-making model (NDM) by the College of Policing UK.

As is evident in the NDM that all decisions made within the police should rotate around the code of ethics. This is so there is consistency and standards of behaviour that law enforcement officers must follow in order to execute their job roles properly and with fairness to the public. Under the policing code of ethics,

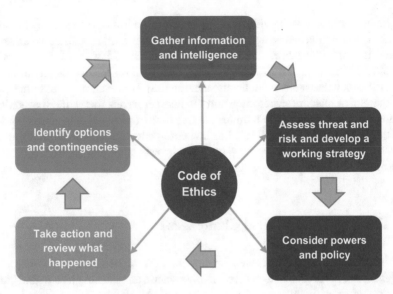

Fig. 1 A visual representation of the national decision-making model (NDM) adapted from the College of Policing UK

the following attributes or 'policing principles' must be adhered to; accountability, honesty, integrity, leadership, objectivity, openness, respect, selflessness. It can be argued that the code of ethics limits room for change and movement by police, however if there was no code of ethics to follow the police would get a lot more complaints around the morality of their actions and behaviour by the public. The police would be largely judged by society as unethical, even if the officers dealing with a situation were not acting unlawfully. However, the code of ethics should be acknowledged on a case-by-case basis not a generic basis and the code updated to keep up with experience [29]. Law enforcement ethical issues arise in all aspects of the job. One of the main ethical issues that will arise is being concurrent with online policing and upholding the law and human rights. It should be considered that it may be difficult for officers to do so where the circumstances are ambiguous. For example, someone who is found with a small number of drugs on their person during a police search, could end up with a jail sentence, face fines or risks to family life and social status, however the police cannot take this into account when arresting this person because they have broken the law, therefore the officer conducting the search would be in breach of the law if they did not arrest.

4.2 Personal Privacy

Personal privacy online is a major ethical debate, as well as being largely subjective. The privacy that each individual has the right to is no argument, however it must be made clear at where the line is drawn into looking into people's personal information. Whether that would be personal data or other data such as search history, data stored on an electrical device or purchases made online. Technology is pretty much linked to every aspect of life, we shop, bank, vote, work and learn via the internet [30]. When people search the web, they should be able to do so without the risk of implications. This is a very subjective matter however, some people will have more lenient opinions on this, and others believe that being able to search for whatever you like via the internet could lead to harmful implications with harmful consequences. Which is not untrue, many undesirable and harmful things can appear on the internet that not everyone would deem appropriate to see. Harmful content online is easy to access, especially through private browsing modes or dark-net sites. Many users will anonymise themselves online in order to access harmful content such as extreme pornography, child images or express unpopular opinions as discussed in Sect. 3.

4.3   Police and Legal Barriers

In England and Wales, anyone who takes part in organised crime is punishable on indictment for 7 years or more. Participation is criminalised under the Serious Crime Act 2015. The definition of organised crime whether that be online or offline is three or more people who exist for a period of time, agree to act and further a criminal purpose. It has been argued that essentially this definition overlaps with other high-risk crimes with high levels of harm in various ways. However, online crime covers a number of separate areas, such as fraud, harassment, illegal purchasing, that may not be specific to online criminal activity and may not reach the threshold for sentencing for organised crime [31].

4.4   The Computer Misuse Act 1990

The legislation governing computer integrity crimes in the UK is the Computer Misuse Act (CMA) 1990, the USA is governed by, The Computer Fraud and Abuse Act 1986 and most of Europe follow the guidance in the Council of Europe's convention on Cybercrime. This legislation protects the offence of unauthorised access to a computer with intent to commit a further offence and the modification of computer mechanisms [32]. Although, within the first decade of CMA 1990 becoming legislation in the UK, there were very few convictions under the Act, only around 100 prosecutions against any cyber criminals [11]. The CMA 1990 has been harshly criticised as not fit for purpose because of the age of the act, digital society has come a long, long way since 1990. Mark Zuckerberg, the founder of Facebook, was just six years old when this Act came into play. It has been argued that the CMA does not reflect on the harm that online criminals cause to their victims. The act has been conflictedly defined as 'flexible' and able to be widely interpreted by the courts, in favour of the Act, but clearly there is room for improvement. This Act has since been amended twice, by the Police and Justice Act 2006 and the Serious Crime Act 2015. The current Act governs:

1. Unauthorised access to computer material.
2. Unauthorised access with intent to commit or facilitate commission of further offences.
3. Unauthorised acts with intent to impair, or with recklessness as to impairing, operation of computer.

   And further offences of:

4. Unauthorised acts causing, or creating risk of, serious damage; and
5. Making, supplying or obtaining articles for use in offences under all four sections above.

The act has been disputed to not keep up with ever changing technology and knowledge as well as the evolving threats that come from the online world. Current society has had a huge increase in extortion via hacking that was reported to Action Fraud in the year 2019, as has no doubt increased during the pandemic in 2020 and 2021 [33]. The main issue here is that there is a genuine difference between using a computer to commit cybercrime and a computer being the main aspect of a cybercrime.

4.5   RIPA 2000

The police and law enforcement agencies are able to obtain communications data of individuals or groups under the Regulation of investigatory Powers Act 2000 (RIPA). This data can be IP data, phone or computer usage for example. It must be however, made sure that communications data is not requested from medical professions such as doctors, members of Parliament, ministers of religion, lawyers or even journalists [29]. The degree of interference with privacy may be higher where the communications data being sought relates to a person who is a member of a profession that handles privileged or otherwise confidential information. The Metropolitan Police and Kent Police were both evaluated after gathering communications from a journalist. Possibly arguing that RIPA is not fit for purpose. Realistically ethical issues should not arise from RIPA as the main form of communications that people reply on are digital communications these days. Therefore, it may be hard to argue that many ethical challenges arise here. RIPA allows the police to access communications data of criminals involved in an open investigation, such as those who have accessed and downloaded child images, extreme pornography or terror related content in order to track down suspects and execute search warrants of their properties to search and seize anything that may be relevant for investigation. Realistically the police being able to gain access to communications data meets the needs of the digital age [29]. This should essentially mean the location of who uploaded the original data may be identified by law enforcement. Issues will arise where botnets are used. Innocent victims may become 'suspects' and go through the trauma of searches and seizures of routers or electronic devices when in fact they are just innocent victims who have had their IP addresses or phone numbers used for crime [4]. This poses a possible ethical issue however, for wrongfully being involved in a crime that these victims did not commit. It could be argued that we need to replace the code of ethics about the digital world to one of its own, that relates only to electronic crime.

4.6 The Dark Web

As mentioned previously, the dark web is full of users who have used technology in order to hide their IP addresses to become anonymous users. The dark web is seen as a space that is out of reach of law enforcement agencies which is not necessarily true. UK law enforcement remain secretive around the techniques that are used in order to catch criminals on the dark web, this is so, inevitably, they can continue to do so without the knowledge of said criminals [4]. Entrapment is an illegal concept that officers will often be accused of; it is where criminals are tricked into committing a crime in order for law enforcement to secure their prosecution. However, the definition covers an 'innocent person' not necessarily a known criminal. Entrapment compromises the integrity of the judicial system, therefore applications that are made in this way are considered made in bad faith and will most likely be thrown out [4]. Therefore 'forum shopping' is seen as a legitimate way to gather intelligence on dark web sites in order to prosecute fairly and lawfully under the police codes of practice above. Going undercover on the dark web is a useful tool for investigators in order to infiltrate forums in order to gain intelligence and have useful surveillance of criminals on the dark web (forum shopping). Officers will infiltrate forums in order to draw out any criminals, for example drug traders or paedophiles.

An example of success by undercover officers was during the Silk Road take down, officers remained on these sites undercover in order to draw out more criminals involved in the trading. Gaining and maintaining trust is the prime factor in undercover operations [4]. Polemically, the extent in which officers go about their undercover work that would be illegal otherwise, is no doubt an ethical argument. Jurisdictionally, many countries or cities may have extensive powers in order to go undercover and police must stick to this, thus, leading to challenges for police. Looking at undercover infiltration in a more legal sense, there is arguments that arise in regard to this because of evidential difficulties. Under UK law, should it be legal to obtain evidence from a course in this manner? No case in the UK has yet been thrown out because of this but it is argued that on the grounds of fairness, under Section 78 of the Police and Criminal Evidence Act (PACE) 1984, there may be the general exclusion of evidence. This is due to police using an illegal site in order to observe criminals being seen as 'unfair under Human Rights laws. This denotes that the court may:

refuse to allow evidence on which the prosecution proposes to rely if it appears to the court that, having regard to all of the circumstances, including the circumstances in which the evidence was obtained, the admission of the evidence would have such an adverse effect on the fairness of the proceedings that the court ought not to admit it [34].

5 Possible Avenues for Future Research

As discussed in the previous sections, there are many technical, non-technical, legal and ethical challenges that police face in policing cyberspace and the dark web. This section will dive into the future avenues for research into policing cyberspace and look at the ways in which the police are effectively adapting to the digital age. The policing of cybercrime needs a wide-ranging development and presentation of legal measures and technical measures to refer to any issues that come about. This section will touch on international coordination requirements, police techniques, the need for increased digital literacy and more future possibilities for the police and private sector.

5.1 International Networks

As mentioned previously, having good international networks plays a huge part in being able to police cyberspace effectively. The police will need strong international networks to enable data sharing between governments [35]. We need international harmonization in order to successfully prosecute suspects of cybercrime globally without setbacks. Transnational and 'borderless' cyberspace lead to prosecution difficulties revolving around cultural and religious differences all around the globe. Religious online offences sometimes make it impossible to prosecute cross-border. There is opportunity for stronger networks internationally and more co-operation across borders for different units in different countries to investigate, not cybercrime as a whole, but on a case-by-case basis. Meaning that generally, as more and more cases are investigated by trans-national police units, a greater understanding of how to deal with certain issues will come to light. This means that countries or areas with less reach that others will be able to get guidance and help from countries that have that capacity, leading to higher success in prosecution of online crimes and more intelligence of crime in cyberspace.

5.2 Increased Digital Literacy

In order for success rates for prosecution from investigations to increase and the difficulty that police face every day in police cyberspace to decrease, there needs to be a greater understanding from both the police and the public in regard to the internet and the online world. General digital literacy in the police will need to improve. The demand for new skills is imminent. Techniques in policing the real world and to then maintaining and policing cyberspace leaves implementing policy concepts or reviewing old ones, such as justification of a breach of public order or human rights, as a priority. Tatiana Tropina has stated 'On the global level police initiatives are represented by the activity of Interpol, which regards the fight against cybercrime as a part of a global security initiative, which includes computer forensic, online investigation, training, public–private partnership, review and evaluation of technology and law enforcement. As part of the agenda for the creation of effective mechanisms for policing cyberspace, Interpol intends to operate both on the level of regional working parties and on the global level, facilitating sharing information within participants' [15].

Because of the ever-growing nature of cybercrime, more funding has been put into police forces that are specialised in the investigation of cybercrime and policing cyberspace. This funding will ultimately lead to more experienced and better fitting police staff to work in these specialised units. Being able to put these police staff through more specific courses or more expensive courses to better qualify and for staff to increase knowledge on cyberspace and how to investigate certain situations. Consequentially, this will lead to better literacy of cybercrime and cyberspace within the police. As well as the police needing increased digital literacy, it is also important that the public have a better understanding of cybersecurity and how to protect themselves online. Investigators with high workloads will not always have the time to deal with such cases that may not be able to be investigated, due to not being able to go over the heads of companies such as Facebook, Instagram and Twitter because of their American heritage. However, all policing units take the time to do so, giving protect and prevent advice to the public. Nonetheless, this advice will come as a result of a crime being reported, which almost appears as too late for the victim. Police should give frequent talks or have advice helplines in order for the public to know how to properly secure their digital devices, before the crime occurs. This will ultimately make life a lot easier for the police as they will be spending less time giving protect advice and advising victims that there are no investigatory means or proportionate lines of enquiry rather than taking this time to deal with high-risk online crimes and following investigatory pathways.

5.3   Open Source Intelligence

One of the most used techniques to police the dark web that is widely known is 'Open-Source Intelligence' or 'OSINT'. OSINT is data that is recorded publicly and is easily accessible to policing bodies. The means of retrieving this information is legal and requires little effort, making it a desirable technique for police to use. OSINT is not limited to the dark web but can be used to retrieve intelligence for any investigation including cybercrimes. Usually leading investigative bodies to small pieces of information that are left on the web that could have been accidental or in human error. Examples of OSINT sources include chat rooms, web pages, media sources or forum posts. However, OSINIT has been criticised for its inadequate abilities in recording OSINT [4]. Though, cyber criminals especially have a thin line in which masking their identity online is valuable to them and it being invaluable to their sales. A criminal will want to make himself known online in order to gain customers or knowledge of his online trades and make profit from it. It is possible that these thin lines that criminals face may leave room for OSINT to do its magic. OSINT was the most valuable tool used in the famous Silk Road trade takedown. A personal email address was linked to the Bitcoin forum that it was advertised as originally, leading police to the criminal source [4]. The same happened when the director of 'Cali Connect' tried to trademark his online drugs trade under his own name, again leading law enforcement agencies straight to him. Generally following the flow of Bitcoin is the most successful technique that police investigators have used to catch criminals online. Another is, when online criminals trade, whether it be weapons, drugs or anything physical, they need means of transportation and delivery. This is usually where human error will occur and police will intercept, following money, transfers and people delivering the goods in order to catch the criminals involved [4].

Public policing is not the only way in which the online world can be policed. It has been argued that 'the public are the police' in regard to online safety and crime. The twenty-first century allows for expectations that are not met by the police to be implemented by the public [35]. The line is drawn by the public as to what is acceptable and what is not. A possibility could be that an in between policing body, that is not part of a force but a voluntary service that people with cyber abilities can sign up to in order to teach or spread awareness of cyber safety and digital technology. Almost like a good Samaritan service. An example of public interference that assisted police and ultimately ended with a suspect arrest, is displayed in Netflix series 'Dont Fuck With Cats'. The immense power of the public is shown in this series. Information is already widely available online however not easy to navigate if you aren't technology savvy, so it may be easier to learn from a tutor. This will help members of society to learn how to protect themselves online and allow for policing bodies to spend their time looking into high risk or more serious cybercrime rather than simple compromises of accounts [36]. High tech cybercrime centres such as the European Cybercrime Platform created in 2008 allowed for Europol to work together with private sector organisations to build programs to train on fighting cybercrime, these included AGIS, ISEC and Falcone which are used to increase capacity within the police and allows for smooth running cybercrime training among different police forces [37].

The harmonisation of police and private partnerships has been argued as the greatest auspicious ways in which the future of policing online crime and cyberspace depends on [15]. Private companies have played a huge part in the development in fighting cybercrime and ICT innovation. The private sector, unlike the public sector, has more of a covert understanding of networking and criminal activity online as well as ICT technology in general. The constant changes in technology and society means that it is more difficult for the police to keep up with these changes and having certain units in the private sector to research and constantly train to keep up with this demand. The police and the private sector complement each other nicely and lead to the cooperation from both private and public sectors. This cooperation means that there is better perception of cybercrime and allows the issues at hand to be tackled and prevented by the police with the help from the private sector and vice versa [36]. The enhanced knowledge of private sector experts is required by the police because of the lack of funding and specialists in the public sector, mainly because of the massive difference in pay scales for cyber–professionals in the public sector compared to the private.

5.4 A Social Media Presence by Police and Threat of Regulation

Regulation of social media companies have been argued to have been insufficient for the prevention and detection of crime online. Therefore, the government has put in place regulatory codes of practice under the Online Harms White Paper, where a duty of care must be complied with on all online platforms [38]. Independent regulators will make sure that the duty of care is complied with by these companies, and they will promote education and awareness around online safety. Legal implications from not complying with these codes of practice will result in guidance and advice, removal orders, fines, disruption of activity and lead to ISPS blocking. These implications are in place to ensure that companies comply with the paper and online harm is kept to a minimum. Companies that run online platforms are expected to go above and beyond in order to comply with these statutory duties from the government. Although, most social media giants such as Facebook or Twitter are self-regulating, meaning that built into their online platforms there will be software which will be able to analyse language, remove 'clusters' of malicious actors and place digital fingerprints on accounts that have posted harmful content previously in order to stop them from doing so again [39]. This content will consequentially be removed from the site automatically, without being done by a single person. Issues still arise however, concerning harms that arise on social media platforms, maybe more privately, with respect to direct messages and in relation to unauthorised access to accounts. This can lead to difficulties, mainly to do with jurisdictional issues as mentioned previously. Most social media giants have USA origin, meaning that the UK has no jurisdictional hold over them. Investigators in UK forces cannot contact these companies directly to gather information of unauthorised access nor to ask them to delete or report an account with suspicious activity. This is up to the victim to do themselves, however these companies are under no obligation to reply at all, and most victims are left with open accounts that have been taken over, that will probably never be closed, but they have no access to anymore. This is one of the major issues that stems from petty online crime, especially here in the UK due to most major companies for social media originating from the USA.

6   Conclusion

This chapter has discussed technical and non-technical challenges of policing cyberspace as well as the legal and ethical challenges that follow. The chapter has also discussed possible future avenues in which police should follow in order to successfully police cyberspace in the future. Whilst exploring the different non-technical issues that came to light, it has been made clear that having an online presence is easy and desirable and has become a major aspect of modern society. This ultimately means that the majority of the population have an online persona and the consequences of cybercrimes must be taken more seriously. The global reach that cyberspace allows for cybercrime means that anyone who has access to the internet via a digital device may be a victim of cybercrime at some point in their life. This means that jurisdictional issues arise when looking to prosecute cross-border. Day-to-day crime has been the instigator of cybercrime as criminals find it more appealing to go about their criminal activities online rather than physically. Police struggle with a lack of resources to chase every crime and investigations may come to a standstill causing victims stress and cause the government financial loss. Cyber units encounter issues where cybercrime is under-reported mainly because of the publics narrow trust in the police but also their lack of understanding of cybercrime.

When diving into the technical issues the police face when policing cyberspace, it has been highlighted that anonymity from cyber criminals online is one of the biggest hurdles that police are faced with. Generally, those who browse the web for legitimate reasons do not need to anonymise their server, those who are using the web for illegitimate reasons, will choose to do so. Although, some users chose to anonymise themselves online using VPNs because they can and feel that it is a basic right to search what they wish without implications. The resources of cyber units within the police do not always stretch as far as specialist technology to collect data from anonymous sources to aid investigations and they struggle as a result. Cryptocurrency is the new most popular way of investing money virtually. Crypto wallets can be used to store money online and make payments that do not come from a registered bank account, meaning it is the most desirable method of payment for dark web users and online scammers. Cyber-criminals will use it to transfer money for illegal trade and scammers will use crypto wallets to mask their identity when tricking vulnerable internet users to send them funds. The police code of ethics is the basis in which all officers and investigators must follow when gong about daily tasks. Police use the National Decision-making Model (NDM) to make decisions regarding all different types of policing scenarios. The same goes for those investigating online crime. Personal privacy is a major ethical argument amongst policing bodies as it is a difficult weighing out human rights and law. It has been argued that the Computer Misuse Act (CMA) 1990 is unfit for purpose due to its age and inability to keep up with ever-changing technology and society. The act may need to be amended again soon, or completely re-written to fit today's modern age and the new age of cyber criminals. The dark web itself leads to ethical concerns, this is because using the dark web is in no way illegal but fear or looked down on by thousands and may even be used to implicate a person of interest in an investigation if they are a dark web user as opposed to a user who is not. Police infiltration via the dark web, of forums and blogs may also raise ethical conflict due to dodging entrapment laws.

Some of the future avenues for policing cyberspace have been explored. Going forward, international networks must be stronger. One of the most difficult obstacles to overcome as a police investigator is jurisdiction and where you can and can't move freely with the law, what you can and cant look at via the internet when investigating a suspect and always having rationale for doing so if you do. The endless reach of cyberspace creates massive issues when trying to investigate and prosecute people of interest overseas because of legal and cultural differences globally. Law enforcement agencies need to increase the digital literacy, not only within policing bodies but also with the general public. More specific courses and internal training would be beneficial for police units and more education and awareness of cybercrime, digital environments and online security would be beneficial for the wider public. The public are an important factor in regard to policing the online world. As well as the public, the private sector also play a vital role in policing cyberspace because of their increased knowledge and specific skillset, both the public sector and the private sector complement each other perfectly. The police have regulatory codes of practice on social media platforms, that these companies must comply with. Meaning that less harmful content is exposed via social media platforms such as Facebook, Instagram and Twitter, as well as these giants having their own regulatory software such as language identification. These future avenues will be advantageous for law enforcement when policing the digital world.

References

1. Bussell J (2021) Britanica. https://www.britannica.com/topic/cyberspace
2. Gibson W (1982) Neuromancer
3. Davis C (2017) Addressing the challenges of enforcing the law on the dark web. GlobalJus-ticeBlog.com, pp 1–2
4. Davies G (2020) Shining a light on the policing of the dark web: an analysis of UK investigatory powers. J Crim Law
5. InternetMatters (2021) What is the dark web? https://www.internetmatters.org/hub/guidance/what-is-the-dark-web-advice-for-parents/?gclid=EAIaIQobChMI9efz98-D9AIVBeh3Ch3FKw2OEAAYBCAAEgJnh_D_BwE
6. Russell T (2021) Silk Road review: the true story of the dark web's illegal drug market. Read more: https://www.newscientist.com/article/mg24933260-400-silk-road-review-the-true-story-of-the-dark-webs-illegal-drug-market/#ixzz7BR7Nh3xx. Retrieved from New Scientist: https://www.newscientist.com/article/mg24933260-400-silk-road-review-the-true-story-of-the-dark-webs-illegal-drug-market/
7. Tabachnick C (2019) The pitfalls of policing the dark web. Retrieved from World Politics Review: https://www.worldpoliticsreview.com/articles/27136/the-pitfalls-of-policing-the-dark-web
8. Rudd A (2018) Law enforcement crack down on the dark web. In: National Cyber Security Centre annual conference. Home Office, Manchester

9. Leivesley S (2016) Five challenges from cyber space for policing. Retrieved from CIFAS: https://www.cifas.org.uk/insight/fraud-risk-focus-blog/five-challenges-from-cyber-space-for-policing
10. Lewis M (2003) Policing cyberspace. Retrieved from ComputerWeekly.com: https://www.computerweekly.com/feature/Policing-cyberspace
11. Wall D (2011) Situating the public police in networks of security within cyberspace. Policing cybercrimes
12. Morgan S (2020) Cyberwarfare in the C-Suite. Retrieved from Cybercrime magazine: https://cybersecurityventures.com/hackerpocalypse-cybercrime-report-2016/
13. Gross G (2018) The cost of cybercrime. Retrieved from Internet society: https://www.internetsociety.org/blog/2018/02/the-cost-of-cybercrime/?gclid=EAIaIQobChMIzKffoO-u9AIVIYxoCR2KDQ-IEAAYAiAAEgKH5PD_BwE
14. Stalans L (2016) Understanding how the internet facilitates crime and deviance. Int J Evid Based Res Policy Pract
15. Tropina T (2009) Cyber-policing: the role in fighting cybercrime. In: Conference in Bad Hoevedorp. European Police Science and Research Bulletin, Germany
16. Webber M (2001) Jurisdictional issues in cyberspace. Tort Insur J
17. Jumari V (2011) Cyber law. University of Lucknow, Lucknow
18. Wollacott E (2019) Why police need the skills to counter cybercrime. Retrieved from Raconteur: https://www.raconteur.net/leg
19. HomeOffice (2002)
20. McGuire M, Dowling S (2013) Cyber crime: a review of the evidence, chapter 4
21. UNODC (2019) Obstacles to cybercrime investigations. Retrieved from Promoting culture and lawfulness: https://www.unodc.org/e4j/en/cybercrime/module-5/key-issues/obstacles-to-cybercrime-investigations.html
22. Dergisi G (2020) Examining the challenges of policing economic cybercrime in the UK. Research Gate
23. NIJ (2015) NIJ. Retrieved from Taking on the Dark Web: Law Enforcement Experts ID Investigative Needs: https://nij.ojp.gov/topics/articles/taking-dark-web-law-enforcement-experts-id-investigative-needs
24. Young T (2020) a10. Retrieved from What is Carrier-grade NAT (CGN/CGNAT)?: https://www.a10networks.com/blog/what-is-carrier-grade-nat-cgn-cgnat/
25. Hartwig B (2021) Cybersecurity in cryptocurrency: risks to be considered. Retrieved from DATAVERSITY: https://www.dataversity.net/cybersecurity-in-cryptocurrency-risks-to-be-considered/
26. GOV.UK (2018) Data protection. Retrieved from GOV.UK: https://www.gov.uk/data-protection
27. Wall D (2007) Policing cybercrimes: situating the public police in networks of security within cyberspace. Police Pract Res
28. Thomas J (2005) The moral ambiguity of social control in cyberspace: a retro-assessment of the 'golden age' of hacking. New Media and Society
29. Rozenburg J (2015) The Guardian. Retrieved from Police force ethical dilemma in increasing digital world: https://www.theguardian.com/law/guardian-law-blog/2015/jan/12/police-ethics-digital-internet-technology
30. Zhong N, Liu J, Yao Y (2002) In search of the wisdom web. IEEE Comput (IEEE) 35
31. Leukfeldt R (2016) Organised crime group consists of three or more persons who act, or agree to act, together to further a criminal purpose. Eur J Crim Policy Res
32. Gov (1990) Gov.uk. Retrieved from legislation.gov.uk: https://www.legislation.gov.uk/ukpga/1990/18/contents
33. Yapp P (2020) The 30-year-old Computer Misuse Act is not fit for purpose. Retrieved from Tech Law for everyone: https://www.scl.org/articles/10854-the-30-year-old-computer-misuse-act-is-not-fit-for-purpose
34. Police and Criminal Evidence Act (1984)

35. Dick C (2021) The future of policing in the digital age. Retrieved from DITCHLEY: https://www.ditchley.com/programme/past-events/2010-2019/2018/future-policing-digital-age
36. LMCC (2003) Digest of electronic commerce policy and regulation
37. OSCE (2008) Fighting the threat of cyber crime. Report Annual Police Experts Meeting, Vienna
38. THF (2019) Our submission to the Department of Culture, Media and Sport and Home Office consultation on the Online Harms White Paper. Retrieved from The Health Foundation: https://www.health.org.uk/news-and-comment/consultation-responses/online-harms?gclid=EAIaIQobChMIs5uOx9zg9AIVHertCh0CYAoQEAAYAiAAEgJnGPD_BwE
39. Bishop P (2020) Lecture 5 cybercrime in cyberspace. Swansea

Contemporary Issues in Child Protection: Police Use of Artificial Intelligence for Online Child Protection in the UK

Christantus Tabi, Chaminda Hewage, Sheikh Tahir Bakhsh, and Elochukwu Ukwandu

Abstract This chapter investigates Artificial Intelligence (AI) inspired approaches used by the police in protecting children online. The reviewed approaches are successful in most of the situations but have their own weaknesses. As such consideration is required for all stakeholders within the child protection arena. The utmost duty to protect children lies with all, irrespective of whether the abuse occurred on or offline. The reporting and intervention on child abuse cases were based on the community, as this was mostly offline perpetrated by parents or caregivers. However, with the advent of technology and the increasing use of the internet by children for several reasons, it has shifted most abuses from offline to online. The law enforcement authorities such as police plays a vital role in protecting children online and can apply different approaches compared to other agencies such as Social Services, Health, and Education. However, Government recommendations for a joint working response mean that all child-protected agencies need to work together in the process of protecting children (HM Government in Working together to safeguard children: a guide to inter-agency working to safeguard and promote the welfare of children, Department for Children, Schools, and Families, London, 2010). However, with the emergence of COVID-19 and the high reliance on the internet by children, it meant that the police must adapt to the changes and rely on advanced technologies such as AI. The UK Police force is stretched due to a lack of financial and human resources, which means that alternative intervention methods are applied in monitoring and attacking online child abuse. This chapter challenges the use of AI

C. Tabi · C. Hewage (✉) · S. T. Bakhsh · E. Ukwandu
Cardiff School of Technologies, Cardiff Metropolitan University, Western Avenue, Cardiff CF5 2YB, UK
e-mail: chewage@cardiffmet.ac.uk

C. Tabi
e-mail: ctabi@cardiffmet.ac.uk

S. T. Bakhsh
e-mail: sbakhsh@cardiffmet.ac.uk

E. Ukwandu
e-mail: eaukwandu@cardiffmet.ac.uk

© The Author(s), under exclusive license to Springer Nature Switzerland AG 2023
R. Montasari et al. (eds.), *Digital Transformation in Policing: The Promise, Perils and Solutions*, Advanced Sciences and Technologies for Security Applications, https://doi.org/10.1007/978-3-031-09691-4_5

unilaterally in predicting and identifying online abuse as opposed to face-to-face investigation and intervention. Though AI can be helpful, it has limitations that can impact on protecting children online as discussed in this chapter.

Keywords Child protection · Artificial intelligence · Big data · Child safety · Online abuse

1 Introduction

There is a wide-ranging consensus that internet usage during COVID-19 was in an increase compared to pre-COVID-19. The situation post COVID-19 is still unknown, as it is uncertain whether internet dependency will continue as is the case during lockdown. Internet use by children and others was already in the increase, but COVID-19 speed up the process. For now, most of the statistics on the internet usage is based on pre and during lockdown. The impact is felt around the world by internet consumers, providers and others respectively [75]. Maria Rua Aguete of Omdia, reported that "increased consumption of digital content from mobile apps to free TV streaming and gaming has already been observed in China and Italy" (www.Omdia.com). However, it is also acknowledged decreases in revenue for some industries such as in the creative industries and TV advertiser when payments are required and are requesting French government intervention in the case of France. While countries like Argentina, Chile and Peru are applying different protection measures such as social security and tax concessions and emergency payments to help relief sectors affected by the impact of COVID-19 on their media sectors [2].

It can be argued that, if lockdown was a pull factor for reliance on internet services, then once lockdown is completely removed, the demand and use of internet would be expected to decline significantly.

One of the COVID-19 legacies is that interventions and support are now gradually moving from in-person interaction to digital and online service provision and intervention. The risk identified for children whilst massively depending on online for several reasons and spending huge amount of their time on the screen is acknowledged in many domains. These resulting effects of the pandemic suggest that the use of digital technologies by the law enforcement officers in policing online crime globally is inevitable.

1.1 Big Data Uses

There is no generally accepted definition for big data. According to De Mauro et al. [14] "Big Data is the Information asset characterized by such a High Volume, Velocity, and Variety to require specific Technology and Analytical Methods for its transformation into Value" [14]. The understanding of this definition is that for materials to be considered as 'big data', they must be massive, easy to access and use for the required purpose. However, this is not always the case as specific traditional intelligence can be relevant in dismantling crime against a child. This is common in child exploitation, slavery, and county line concerns [48, 76].

Meanwhile, in Europe, the European Commission has an encompassing definition which stated that "large amounts of different types of data produced from various types of sources, such as people, machines or sensors [8]. This data includes climate information, satellite imagery, digital pictures and videos, transition records or GPS signals. Big Data may involve personal data: that is, any information relating to an individual, and can be anything from a name, a photo, an email address, bank details, posts on social networking websites, medical information, or a computer IP address" [25].

Koops, Bert-Jaap has reiterated the importance of bid data used by intelligence agencies in many circumstances and that a key method of providing intelligence has always been to gather large amounts of data from numerous sources. Since the last few decades, even law enforcement has begun to use data mining as a means of gathering information to aid investigations [44] and punish criminals [31, 43].

1.2 AI Use in the Modern World

There are different definitions for Artificial Intelligence. One of the definitions that attract attention and consideration is that provided by the European Commissioner for Human Rights, based on three aspects (data input to perceive the virtual environment, use the perceptions, and derive outcomes) [81]. However, in this paper the definition that would be considered is that provided by Hunt et al. [38]. According to Hunt et al., "Artificial intelligence (AI) refers to the capacity for learning and "intelligence," which can be demonstrated by computers and machines". Though this definition was used in the context of understanding how AI, Big Data, Machine learning, and mHealth are connected in early prevention, detection, and response to violence.

The reliance on AI and big data in the global health domains [85], is an indication that it can play a big role in the fight against violence towards women and children. AI and big data are not only an interest in health but have also drawn interest from law enforcement [53] and other businesses [42]. Key elements in the use of AI and big data resulting from advances in technology are that it provides an opportunity for agencies to access a huge amount of information called data instantly and, in some cases, accurately [61]. The correlation between AI and big data is interesting because for data to be generated, AI is required.

Some years ago, computational approaches of 'predictive crime mapping' began to be used in crime prevention. Predictive policing tools based on "big data" have advanced once again. To begin with, AI technologies promised to make sense of massive volumes of data and extract meaning from disparate data sets. Second, they signaled a change from decision support to primary decision-making. Third, they are intended to regulate society as a whole, not only the fight against crime [74]. AS far as child protection is concerned, the UK police would normally consult their database called the NICHE to check previous police involvement of an offender. The use of image recognition technologies and software such as Apple child sexual abuse materials in the US will be helpful when rollout in the UK and other countries. Apple's idea is paving the way for other tech-companies to develop tools that will enable law enforcement officers to scan child sexual abuses images on phones, but for now it is only possible with iPhones with its Siri iMessage tool. Though this is a step in enabling effective control and deterring offenders from holding and sharing child sexual abuse images, the officer must have access to the device in question. Furthermore, the abusive and offensive images of children store in iCloud must be coordinated alongside National CSAM database to enable quick recognition by the law enforcement.

1.3   What Is Child Protection/Online Abuses

Children's welfare has attracted huge attention since the tragic death of Victoria Climbie and was reinforced after the death of "Baby Peter". Though the emphasis on the failures of agencies was mostly on health and social services, the police have a role as welfare professionals to ensure that children's safety is priorities. The guiding principle for child protection had always been collaborative intervention and was an emphasis in the government guidance known as "Working Together Under the Children Act 1989: A Guide to the Arrangements for Inter-Agency Cooperation for the Protection of Children from Abuse [37], which had subsequent reviews, but with the focus on child protection and what constituted Safeguarding. When it comes to Child Protection, the test stipulated by the Children Act 1989 are, reasonable cause to suspect, likely to suffer harm; already suffered significant harm. The test differs with provisions under S17 of the same Act (S.17 Children Act 1989) which is about prevention and precautionary measures and this goes to corroborate the views of Maslow's Hierarchy of Needs."

In this chapter, the growing concerns around children's activities online are highlighted and we aim at looking at how this can be aggravated if there is an increase in the population of children up to the age of 18 years old. According to statistics from the Office for National Statistics (ONS), the UK population is estimated to be close to above 67 million in 2019, and by 2029 it is expected to be around 70 million and most probably above 71million in 2039. These populations comprise both adults and children (the UK refers to the four countries: England, Wales Scotland, and Northern Ireland). Though this populations increase might be by growth in immigration, this might be different with the impact of COVID-19 as traveling and isolations measures affected the movement of people in and out of the UK. This chapter is more interested in the number of children (0–18 years old).

Information provided by the ONS indicates that while the over 18s have the highest population compared to under 18s, the females under 18s old are more than the males across all the regions and this explains which under 18s population that is likely to be at risk of online abuse. This will be the trend for the next two decades as forecast by ONS. Research conducted by NSPCC with 2.275 children between 11 and 17 old, around sexual abuse indicates that 1 in 20 has suffered sexual abuse in the UK [58]. This represents about 114 children in the cohort study. It might not be a true representation of the figure base on the fact that not all sexual abuses are reported, and these figures are for children who already suffered sexual abuses and not based on the likelihood of suffering sexual and other abuses.

1.4 Government Policies

Most governments, especially the UK government, acknowledged and recognised that children, children are abused online on a regular basis and it is in an increase, as many children are accessing the internet for several reasons, which this paper will not investigate now. According to NSPCC, "Online child abuse crimes have surged by three-quarters and more must be done to protect children online". The number of online child abuse alone is almost 80% between 2016 and 2021, and this figure will continue to increase, if the government does not have an appropriate response. To response to these concerns for children and their families, the UK government has introduced an Online Safety Bill, though with some limitations. The concerns around children's online harm continue to be a central debate for many politicians and others. According to Sophie Sanders who works for the National Statistics Centre for Crime and Justice, though children were communicating online more with people they know pre-COVID 19. She reiterated that "being online can provide great benefits to children now more than ever, but it can also offer major concerns." We can see that most youngsters aged 10–15 years only spoke to people online that they already knew in person, based on data acquired before the COVID-19 pandemic.

However, one out of every six children communicated with someone they had never met in person, and 5% met together with someone they had only spoken with online. Although these conditions may not always result in harm, it is crucial to remember that they all pose significant hazards to children [62].

1.5 How Often Do the UK Police Use Information from the Internet for Policing (Police Intelligence) and for Child Protection Purposes

The importance of the use of digital technologies in crime investigation and punishment can not be undermine. Law enforcement organisations and authorities such as the Interpol rely on digital technologies such as AI and big data in investigating cyber sexual crimes against children. Though, technology is used to combat online crime, it is also helping child sexual abusers, to access and distribute online child sexual abuse materials (CSAM) (UK National Crime Agency). Efforts are made at the European and internationally levels to collaborate in fighting and prosecution offenders of online child abuses. This is reinforced by members states though little is achieved by legislation. The new approached has been through ground-breaking digital technologies and capacities, which permits the gathering, creation and storing of large amount of data in a timely and less costly manner, that could be use as police data.

The law enforcement officers have not hesitated in using information gathered through digital technologies in detecting online crimes against children. Beside the police detective teams using digital technologies in identifying and stopping crime, some NGO's such as End Violence against Children, Thorn and the IWF have applied technical tactics such as AI, hashing and blocks to ensure CSAM circulation is minimized [23]. This approach and tactics used by IWF and the law enforcement agents is effective when supported by human intelligence [39]. Information provided by individual informants directly or indirectly to the police by sometimes store in evidence.com and becomes a relevant data base for the law enforcement agencies.

2 Rising Demand for Police Tasks and Child Protection

2.1 The Core Duties of the Police

The police worldwide are always solicited for several reasons. According to one police officer in the USA, "Officers are expected to be social workers, mental health professionals and several other titles in order to fix the problem of the citizens who call. More is expected with less ability to assist—Patrol officer with five years of experience in the Seattle and Tacoma police departments" [24]. In the UK, the police have designated officers to deal with child protection and safeguarding issues. The officers responsible for child protection matters are within the Public Protection Unit known as PPU and the Child Abuse Investigation Unit, usually called CAIU. In sexual assault or rape cases on a child, the police will use technology in gathering and preserving forensic evidence that could be used in court to prosecute criminals. To ensure that forensic evidence is preserved during transportation to the Sexual Assault Referral Centre (SARC) and in the examination, the police must be part of the process and gathered required evidence in supporting Social Service actions and possible court action [62–67].

Most of the risk children encounter today are online, and it is reported by the Internet Watch Foundation (IWF) that as much as 8.8 million attempts were made by people who wanted to access child sexual abuse content in a period of one month, while the National Crime Agency think that more than 300,000 people in the United Kingdom pose a threat sexually to children either directly or via the internet. As such police presence online is required now more than ever. As a result, police have some interference mechanisms such as the Sexual Risk Orders (2014) to obstruct a perpetrator's access and use of the internet. This is only helpful if there is an identified sexual perpetrator, and a court order is granted to the police. However, the huge number of children sharing images and information online either to known or unknown individuals, suggest that the police are overwhelmed and unable to cope with the speed, considering that police numbers are not increasing proportionately to the number of children using the internet and exposed to online abuses daily. In spite the UK government promise in 2019 to drastically increase the number of police officers, there had been a slow increase since then (155,000 in 2003; 150,000 in 2017; 172,000 in 2010 and 160,000 in 2021) [26], but it remains the lowest compared to other countries in Europe.

The police have a duty to make a referral to the National Referral Mechanism (NRM) if requested by a parent who has concerns around the frequency of a child going missing. It is an escalation mechanism with statutory powers to respond to the sexual exploitation of children. Just like the mechanism under s47 of the Children's Act 1989, this is required in obligating services to share vital information in their keeping concern a child at risk of trafficking. Though the UK police force has a specialist unit to deal with child protection issues/concerns, some other police units such as the British Transport Police (BTP) have a duty to police by acquiring footage from CCTV images and sharing with the investigatory police force to fight against crime on children using the railways.

2.2 Police Duties in Child Protection

"The main role of the police is to uphold the law, prevent crime and disorder and protect citizens. Children, like all citizens, have the right to the full protection offered by the criminal law". Their direct responsibility and duty to protect children at risk of abuse was cemented by Lord Laming's report after the death of Victoria Climbie (2003) and stated clearly that 'the investigation of crimes against children is as important as the investigation of any other serious crime and any suggestions that child protection policing is of lower status than any other form of policing should be eradicated'.

Current crimes against children can be sensitive and require immediate police investigation either as single or joint agency intervention and investigation with Social Services under s47 of the Children Act 1989. Police importance and role within the child protection arena are recognized and given priority over other matters with the creation of the Public Protection Unit (PPU). This police unit is responsible for all child protection investigations regarding reported child abusers or identified incidents of child abuse cases. Generally, all police officers are responsible for protecting children irrespective of whether they are from the PPU.

Police presence in the Child Protection process is unquestionable. Their role has been identified within the Working Together to Safeguard Children (2018). Policy and Multiagency Working to a point where it is impossible to alienate their support and service. An increase in the number of children coming to the attention of police and social services attention, followed by serious case reviews and numerous child protection inquiries in the UK and internationally has helped to validate the need for police involvement in ensuring that children and vulnerable people are getting adequate and prompt intervention and protection. Referrals on concerns around child's safety have also pushed up the number of children coming into the care of the local authorities in the UK. These numbers are outweighing the available carers and places of safety [56].

However, while it is now possible for the police and other child protection professionals to collaborate, it is now possible for child protection professionals to arrive at the best decisions for children at risk of harm or abuse. The police play an important role in protecting and preventing children from different forms of abuse. The police are required in child protection processes to share significant information, they hold about a child and the family. This information and intelligence shared by the police can help and guide professionals within the Child Protection Conference in arriving at an informed decision after understanding the risk of harm. During child protection investigation leading to the child protection conference, the police do not only share information with other agencies, but they also gather information from those agencies involved which would be kept in their police database.

In most countries including Australia, the police are involved in the process of the child protection referral pathway as a mandated reporter and interceptor of child abuse [5]. The responses to concerns about a child are normally in two stages and are proportionate to the degree of identified or perceived risk. The UK law on Children's rights under the umbrella of the Children Act 1989 in England and Social Services and Well-being (Wales) Act 2014. Prior to any child protection investigation, the police are informed and consulted via a professional strategy meeting to share information and determine the level of risk, followed by any need for police investigation should it considered that any criminal offense or the likelihood of a criminal offense against a child has been committed.

The police's role in the immediate protection of a child at risk of harm or neglect is crucial. UK police have been endorsed with the power to remove a child at risk to a place of safety. These emergency powers that can be used by the police, only when it is absolutely required if a child is deemed to be at risk of serious harm or suffering from harm extends to them being able to enter and remove a child without parents' consent. The importance of this power though is hardly used by the police permits assist children services in exercising their duties to protect children from immediate harm or likely to suffer harm (Children Act 2004). This power to remove normally lasts for up to 72 h only, as they are used only when there is a cause to believe that a child is at risk of significant harm or at risk of significant harm if these police protection powers are not put into effect immediately.

Effective Child Protection requires professionals involved, such as the police, social workers, and other participating agencies (school and health services) to follow the stages stated above under S47 inquiries after receiving a referral about concerns for a child. The flow chart above (see Fig. 1) suggests that eight stages should be considered in keeping a child at risk of harm safe. However, the police are required

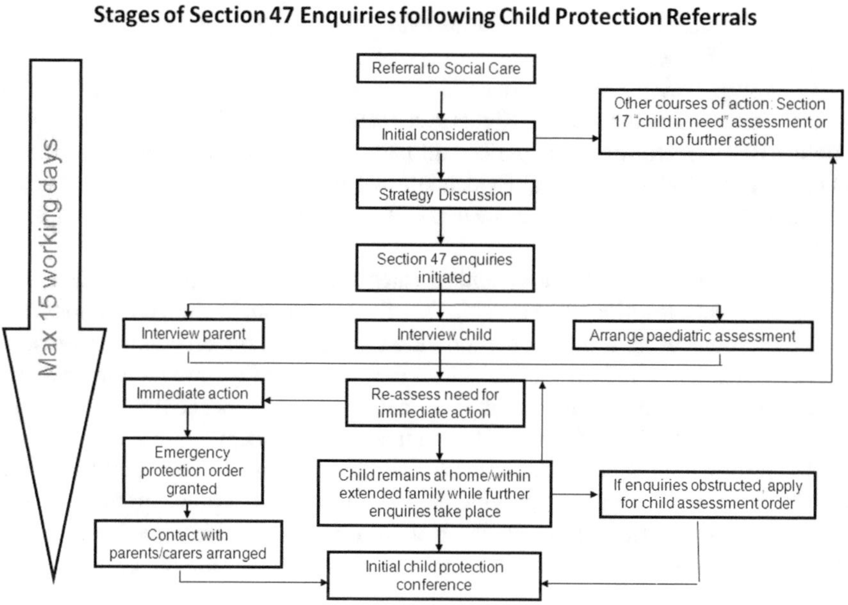

Stages of Section 47 Enquiries following Child Protection Referrals

Fig. 1 Section 47 child protection flow chart [79]

only after the first two stages have been completed by social services. It is relevant to clarify that this process is similar to the process when a vulnerable adult is deemed to be at risk of harm, this seems to be motivated by the fact that the types of abuses that children experience while in the care of other is similar to those categories of abuses vulnerable adults can be subject to, although responses and outcomes can be different for adults [47].

Within its duties in protecting children, the police are required to notify social services and other agencies involved with children of any offense committed towards a child. In the process of protecting a child, the police can

- "take the child away from their home in an emergency—this is called 'placing them under police protection
- share information with the local authority, schools, and health services visit and speak to the child, either with or without parental permission
- visit the home where the child lives or where the offense took place
- search for and seize evidence of the offense
- arrange a medical examination of a child".

According to the Ministry of Justice [54], to enable officers to gather information in Achieving Best Evidence in Criminal Proceedings from children, their age and ability to communicate must be considered when seeing and speaking to them. The acquired evidence is then shared with the courts when required to avoid the child experiencing the stress of court presentation and providing evidence in person. This role by the police is important in securing evidence to enable social services to protect a child from further abuse.

In exerting their powers when a child is at risk, the police are mindful of the right of the family under the Human Rights Act 1998, European Convention on Human Rights (ECHR). However, the police still have rights guaranteed under Section 17 (1) (b), the Police and Criminal Evidence Act 1984 (PACE), and Section 48 of the Children Act 1989 to obtain a warrant that will permit the use of emergency protection power.

The UK police force has a system of categorized information known as POLE (People, objects, locations, and events). POLE is significant in child protection when police are consulting a piece of information from the Police National Computer (PNC). Offenders who have had links with the police usually have their information stored by the police under an incident Unique Reference Number (URN) which would help in identifying the individual if linked to another incident, offense or crime.

The information in police data can be used for enforcing the law. It is also important in the sense that the data provide the possibility for the police to the available information of the offender to predict and analyze possible reoffending. This is where the importance of big data is relevant and vital in protecting children against known offenders. However, this also exposes the gaps in using the police of intelligence in child protection. In the absence of police reports and familiarity with the forces of law and order, though the police can gather data from other agencies such as social services and health for their PND (Police National Database). The police also consult existing database called the NICHE as source of relevant information that can be of help in protecting a child at risk. This information can come from cross-border cooperation, especially when it comes to Registered Sex Offenders within the European community. The aftermath of the impact of Brexit on this cross-border cooperation will not be discussed in this paper, as guidelines and agreements have not been finalized on some aspects of the separation from the EU.

According to UK Government Security Classification (2018), UK police have the power to hold information and protect it appropriately for further use. There are three classifications in order of importance: official, secret, and Top Secret data. The information is based on three principles (All, Everyone, and Sensitive information) All the information that is held for security reasons must be protected adequately. Meanwhile, everyone in contact with this data has a duty to maintain confidentially knowingly or unknowingly, and finally, all the information held on the security database is considered sensitive and only granted to appropriate responsible individuals under full control [6, 12].

3 Police Ill-Equipped for the Role

There is a debate around the police's ability and competencies within child protection services. Just like many sectors, there is a constant reduction of the number of police officers compared to the number of committed crimes, especially domestic abuse.

Though there is no generally accepted definition of domestic abuse and the changing nature of domestic abuses incidents has made it difficult to have a unified definition for domestic abuse. Though there is no legal definition of domestic abuse in the UK, the government and criminal justice agencies identified a wider definition in 2012, to cover most aspects of abuse between individuals of age 16 and over.

Pre-COVD 19, professionals working within the child protection and other sectors of the UK society, were mostly office-based. Child protection and safeguarding officers had to meet and speak to children either at home or in school as a fundamental aspect of the child protection procedure. The situation changed drastically when the COVID-19 pandemic hit the UK and the entire world. The way services had to be delivered, though on a smaller scale, was mainly directed by the national COVID 19 guidelines. COVID-19 impacted the police and other sectors of humanity in all nations in a similar way [77]. The advent of COVID-19 meant a change in normal policing, human activities, and routines were affected in one way or the other. However, it must be understood that this role did not apply to all countries. In the US for example, not all professions were faced with imposed total lockdown [18].

It is imminent that each nation experiences' COVID-19 pandemic on crime is different, while in some countries, reported crime was in an increase, depending on the crime, social distancing measures, and other different preventative actions [55]. Research suggests that while domestic abuse crimes were on the rise during the lockdown [51], other crimes such as urban crimes were lower worldwide. It is also argued that not all states in the US experienced an increase in domestic abuse, and it also varied between ethnic groups and communities too [1]. According to Dai et al. [11] depending on the type of governance, the impact on policing will also be different. For instance, in one of China's provinces due to the very strict application of the COVID 19 lockdown restrictions, the number of police interventions was lower than pre-COVID-19, compared to the state of New York in the US, where restrictions were not compulsory in some professions [11].

To enable effective contribution and collaboration of the police within the child protection arena, police need to be able to assess the information in their keeping, while considering the assessment framework tool. The assessment Framework tool comprises many key domains of a child's life and it does not provide a clear role of the police, though deficiency in any of the domains can enable the police to use their powers to protect by removing the child at the least make an informed decision within the child protection conference with a view of child protection registration. The right to protect children in the UK with the use of the powers to remove a child at risk is endorsed by Sections 44 and 46 of the children Act 1998 to the police, social services, and National Society for the Prevention of Cruelty to Children (NSPCC).

The Assessment Framework focuses on the child's needs and views, therefore in other for the police to capture the views and needs of a child at risk of harm using the assessment framework, they must interview the child and their parents/carers separate base on the child's ability.

4   Rising Population of Children Affected Whilst Online

The COVID 19 brought several challenges as well as opportunities. Children world-wide were affected in one way. Male and females under 18 years were impacted differently [68]. In countries like India [70] and the USA, the experience was not pleasant, because there was an increase in child sexual abuse. Though there is a definition of Child sexual abuse is provided by the World Health Organisation, not all aspects of the definition were prevalent during the COVID 19 pandemic, Due to the inability to have a face to face contact resulting in touching; most of the abuses were via cybersexual acts [87].

5   Data on Children Coming to Police Attention via Online

One of the legacies of Covid-19, is distance contact, and the use of technological devices to gradually replace face-to-face communication. Technology has been rolled out in many domains including education, health appointment, and sports. To enable police officers carryout their work effectively, the law and enforcement departments in most countries have resulted to the use of CCVT, body cameras and intelligence provided by citizens via their personal devices.

5.1   Use of CCTV

There is a constant increase in the number of CCTVs in the UK. This can be explained by its proven importance in controlling crime through surveillance. Although it is difficult to tell how many CCTV cameras are in the UK, there is evidence to suggest that there is a rapid growth in their numbers in all cities. It was commonly referred to as big brother by David Davis, former shadow home secretary in 2008. At this point, he stated that "there is now a CCTV camera for every 14 people" in the UK. It is also claimed that London [49], alone has more than four million cameras and about 4.4 cameras in the UK as a whole by the year 2009 [80].

The use of CCTV by many agencies and the police has been instrumental in moni-toring, controlling, and criminal investigation. In preventing crime, most CCTVs in the UK are linked to the police, but for them to be helpful; they have to be of a good quality image, reliable, and can serve the purpose it is meant for. The images that police obtained from CCTV footage can be used as evidence in court to prove that an offense has been committed. CCTV technology is improving and can clearly link the facial appearance with a known suspect (www.paceinfo.uk). Furthermore, the rele-vancy of CCTV in the UK policing explains why there is an increase in its number [60]. Grabosky [27], states that "CCTV evidence is often very convincing" because the use of CCTV in identifying criminals and controlling crime can be easier. The use of this hi-tech method in monitoring and controlling crime has been efficient and used before the twenty-first century in Australia [27].

CCTV used by the police in child protection, especially when a child is missing regularly is crucial in searching and identifying possible Child Sexual Exploitation (CSE). It is important to note that, the police role is mostly around assessing, managing, and preventing identified risks, but does not hold prosecution powers. The power to prosecute lies on the Crown Prosecution Service (CPS) (www.cps.gov.uk). In helping to keep children safe from sexual exploitation, the police will support parent's request under the sexual Offender Disclosure Scheme, called Sarah's law to have full disclosure of any suspect accessing their child, and can pose a risk to them by virtue of their names been found on the sex offenders register.

5.2 Use of Body Cameras

Though the use of AI and other tools such as CCTV and body cameras are commonly used by the police in gathering relevant data for their investigations, the use of eyewitnesses is also noted to be vital, though there are some doubts when it concerns over 60s, due to possible cognitive issues, when a cognitive interview is not considered [88]. In a study conducted by Kebbell and Milne [41], with 159 UK police officers, they found out that, the police value eyewitness accounts and information in forensic investigation. However, it can be difficult to count on these eyewitnesses' accounts as many of them are reluctant to attend court to give evidence, though the information provided is correct [41]. Therefore, for the police force to effective carryout forensic investigations, it must not only rely on AI, but should also consider information that is available from eyewitnesses.

6 Police Use of AI—Cost-Effective/Mistakes and Challenges

Digital globalization has made countries of the world rely on Artificial Intelligence for some decision-making and speculations. As the digital world is impacting the globe, it is also impacting the way services are delivered and consumed. Therefore, most countries are rushing to embrace the importance of artificial intelligence in shaping its services and decision-making, especially within public services [17, 35]. The police as a public service are not shy in using both "traditional and geo-coded data" in carrying out some of their duties.

The use of mobile devices means that police can gather data easily during the course of their work which can be helpful in keeping the public safe from harm. The use of AI in the form of a computer algorithm is not new to some governments when it is required in decision-making. The level of dependence on automation for public services decision-making is increasing and intensive human involvement is no longer indispensable, though more needs to be done [15, 22, 78].

Henman [36] explored the use of automated decision-making by governments for public services and for AI to be useful, it needs to be operating in three forms such as "detecting patterns: sorting population, and making predictions. Therefore, the use of AI by police can be helpful in safeguarding and protecting children, as available police information generated from AI, would enable the police to detect patterns of social behaviors and abuse online, through the identified populations in records such as the Sex Offenders Register. The possibility of having this information can help in predicting the possibilities of harm towards children while online.

In detecting patterns of behaviors or incidents, AI is indisputably the best way, because the use of superior and bid data processing ability makes prediction more accurate [61] and reliable. In doing so, it will not necessarily require human intervention or assistance. Clark [7] had identified how powerful the use of AI can lead an understanding of human and animal behavior independent of human guidance as was the case of using YouTube Videos to distinguish human and cats faces [7]. However, AI has its limits, as was pointed out in a report conducted by the Medical Academy of Royal Colleges [50]. In this report, it is clear that the validity of AI in the health sector, which should be the case in many sectors that depends on data to analyze and predict possible outcomes [10, 57].

This finding is corroborated in a report conducted in the USA on the use of software called PredPol by the police to predict crime in some towns, by helping reduce the number of police patrol. The research actually found out that the predictions were based on the information or data already fed by the police and did not provide additional value and the number of police patrol was the same and for crimes, that police are already aware of prior to the use of Geolitica (PredPol) crime prediction tool as an AI [52]. The correlations between crime prediction using AI data and the actual arrest of criminals were insignificant as demonstrated in Fig. 2.

Police are employing artificial intelligence (AI) techniques to delve extensively into the planning stages of crimes that have yet to be committed, as well as to examine crimes that have already been committed. Automation tools are supposed to extract plotters of yet-to-be-committed crimes from enormous amounts of data as part of ex-ante preventive efforts. As a result, a distinction is established between tools that target 'risky' persons ('heat lists'—algorithm-generated lists identifying people most likely to commit a crime) and tools that target risky locations ('hot spot policing') [40]. There have been numerous success stories in the battle against human trafficking using the second, ex-post-facto usage of automated systems. To combat child sexual abuse in Europe, Interpol oversees the International Child Sexual Exploitation Image Database (ICSE DB).

Therefore "good-quality AI depends on good-quality data" while good-quality data can only be generated by good quality AI tools. In most developed countries such as the UK, the use of AI is common in the health sector to predict health issues such as mental health resulting from violence against women and children. The fight against violence against women and children is taken very seriously by the UK police and most of the child protection concerns are generated through domestic

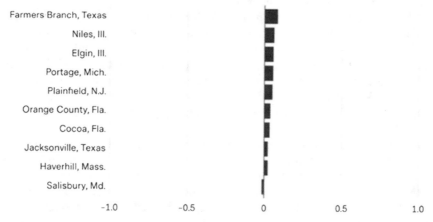

Correlations between predictions and arrests were weak

Correlation between average number of arrests and average number of predictions Graphic: Gizmodo/The Markup. PredPol. police departments for the listed jurisdictions

Fig. 2 Crime prediction software promised to be free of biases. New data shows it perpetuates them—the markup [82]

violence and child abuse. Some other research on preventing crime with the help of computer vision and pattern recognition with supervised machine learning seems outright dangerous [89].

Police assistance in cross-border cases of child protection is always instrumental as they will be able to monitor the entry and exit of children who are at risk of harm or are on the child protection register. This role is always in conjunction with the request of a social worker who is working with a child that is linked to another country. The power to stop and search alongside government guidance in working with foreign authorities [16]. Most law enforcement agencies (LEAs) are using AI to gather; analyze and process data on many aspects such as in social media surveillance [4, 69, 73, 84].

The police are aware that most of the child sexual abuse videos and images are stored in electronic devices such as laptops and telephones. Online gaming has become a new vehicle in sharing child abuse images. As many as 13 million child abuse images are held by the UK Child Abuse Image Database (CAID) and the number is on the increase every month. It has come to the notice of the police through online monitoring that, there is an increase in demand for child sexual abuse videos and images.

According to the Internet Watch Foundation (IWF) in 2018, more than 105,047 URLs and 477,595 web pages were removed due to hosting images and content on child abuse. However, it is interesting to understand that almost 100% of hosted criminal content was not found in the UK, but from what IWF calls INHOPE country. Between, 2019 to 202 there had been an increase of 16% in the number of child sexual abuse images from 2019. IWF report of 2021 identified that about 153.383 images of child sexual abuse were found online.

The use of big data within the police force has helped to shape police relationships with some communities in the US and the UK, especially when it is audio-visual. The use of some high-tech devices to monitor crime has come under scrutiny because the conception is that it can impact some civil liberties. Due to increasing public interest and a large amount of data that can be found easily in several domains such as Twitter, Facebook, MySpace, Tumblr, and other social media platforms. (Politic, psychology, economic and culture) the use of scientific methods by agents of protection is almost irrelevant [3, 83].

When big data is used by the police and other agencies to gather information there is the fear of lack of transparency. The information that the police have either through intelligence or as a result of their intervention is usually obtained without the full consent of the provider, which can pose issues around transparency. However, the issues around accountability and transparency are inevitable when exposed digitally via social media.

The complexity for the police in using available data, which is generally called internet open sources [44], because the information is readily available in the public domain and agencies can be tempted to access and use without due consideration to the rights of the owner. However, the police's powers to use private and public internet sources in criminal investigations are limited by law [19]. Therefore, though the information is in the public domain, it still attracts privacy rights as had been the cases dealt with by the European Court of Human Rights [20].

The situation can be different and more relaxed when it comes to cross-border access to open-sources data as stipulated by The Convention on Cybercrime [9] which is ratified by at least forty countries to help facilitate cross border crime management in accordance with agreed principles by all members in respect to sharing data and information on cross-border crimes. It should be noted that, though there is room to facilitate the use of available data in combatting cross border crimes, Koops highlights that it is usually disputed in spite of the benefits involved [44]. Koops, clarify the use of open-source data by the police by stating that "The Cybercrime Convention does not regulate open-source data collection by the police as far as the collection remains in the police's territory. For cross-border data in open sources, the Convention allows the police to access—and presumably also copy—these without specific authorisation" [44].

> Big data, as we move forward, is going to be one of the most important issues with respect to transparency and accountability.—J. Scott Thomson, Chief, Camden Police Department [24].

The use of AI in policing can be complex and costly as much reliable software is expensive to buy, installed, and renew subscriptions. To effectively carry out some of its duties the police will need cooperation from other public services agencies.

6.1 Challenges and Mistakes

Though the use of big data can be useful in predicting crime and subsequently help in identifying child abuses, it has some limitations. Most of the data that is generated via social media, rather than from a scientific method, indicates that it can be biased [46] and present some mistakes in achieving the desired outcome for the children that needs protection from the police. According to Hargittai [33], the use of big data assumes that most people have used specifically identified social media domains that have been used as a point of references, which can produce errors as not all those who will need police intervention are found likely to be found on the sites that the data was generated for police use. Furthermore, the users of a particular domain and demographic, might not be a true reflection of the age group that requires child protection due to their age, education, and active online use [32].

Another challenge is the reliance on particular social network domains in gathering information, which might not be the only means of communication that individuals or a child are using in communicating online. The other domains which are not regularly used but relevant in communication, such as phone calls, messenger, and face-to-face exchanges), are what Hargittai called the "week ties" [33], and are crucial in reducing reliance on social behavior identified on one site. The use of particular data by the police from specific sites or devices has its limitation. Internet users have different priorities and my not necessary use the same site or social media domain for all their activities, therefore, while they use Facebook for something else, they might be using Twitter for some other things, therefore the dependence of one site to predict social behaviors in preventing crime online could be ineffective [21, 34] (Hargittai and Hsieh 2010b).

The effectiveness of using data to predict crime by the police is a challenge, as data must be at their disposal to enable them to act promptly. This has been a suggestion amongst the medical core where emergency intervention is required [30]. Considering that the police sometimes need to act in an emergency to keep children safe, information about possible offenders needs to be made available to the police force responsible for safeguarding and protecting children.

In as much as the police in the UK, EU and the USA would like to use big data in their various activities that concern the safeguarding and protection of minors, they are not given a free ride as far as data regarding minors is a concern. Data protection has gained some protective status in many countries, as such most countries and regions now exercise data protection laws which can be a hindrance for the police in using available and massive data in policing within the child protection arena. The current information privacy laws in the continents mentioned above are not up to speed with technological advancement and modernization, as such sensitive information concerning minors is still not protected enough [71].

Technology provides opportunities to track down criminals but also poses some difficulties to the police due to constant improvement in smart technologies known as the Internet of Things (IoT). The police are not always capable of keeping pace with the intensification of the Internet of Things, because it is difficult to forecast its development. Nonetheless, reports suggest that by 2030, about 125 billion connected IoT devices will be available worldwide compared to 27 billion in 2017, making an increase of more than 11%. Meanwhile, the data that will be transmitted will also increase from 20–25% to about 50% every year in the next 15 decades [28].

The use of IoT does not only allow the effective and quick collection and sharing of data, but it also becomes cheaper with the introduction of 5 G connections and allows the collection of huge and accurate data, which can be difficult for the police to monitor. Technology and social media facilitate the perpetration of domestic abuse and this can impact police abilities to effectively protect children and women from online abuse promptly. The threats of abuse are high when using technology; this is commonly known as tech abuse. Although technology is easily used to abuse victims, it can also be a reliable and useful tool in preventing further abuses.

This view has been acknowledged during COVID 19, as lockdown and social distance meant that face-to-face was limited and the main point of contact was via technology and internet devices. As more and more people had no easy option to communicate, except via using technology for whatever reason, this increased the risk of online abuse although no hard evidence is available to suggest that the main cause of any abuse during this pandemic was consequential and resulted from the impact of COVID 19.

The use of technology in all aspects of life has its own challenges and errors are bound to exist, which can mislead the intended purpose. In the world of technology, there still exist safety problems that are not technologically related since it is not used appropriately [59]. The magnitude of the issues involved in the safe use of technology is more around the software rather than the technology itself [86].

According to Robert [72], safe use of the technology to assist in all aspects and industries including health care would need to follow some principles. They went further to explore some of these principles such as "*safety is a system problem, not a software or IT problem; safety and reliability are not only different properties, they are sometimes conflicting, and safety must be built into a system from the beginning' it cannot be added to a completed designed or tested into a system*" (2008). This view is relevant in understanding the way information used by the police to monitor and keep children safe online can be controversial because safety will depend on the context and environment.

Therefore, the fact that the police computers are built initially for general use rather than just for policing, makes it flawed and unreliable when software is installed to help in the monitoring. Software are not always a safe way to monitor risk online as they have a life cycle and must be updated regularly to ensure that it can keep to the speed of offenses that are occurring online. When required safety tools and software are not considered at the start of developing a system [45], it can be ineffective in achieving the purpose for which it is intended.

7 Conclusion

The police have a significant role in protecting everyone in the country. However, their duties to protect children from abuse either online or offline are crucial and demanding to the extent that several information sources are required to enable them to accomplish these tasks. And the prevention of feasible crime against children is even more urgent than any other police activity. According to the college of policing, UK "collection, accurate assessment and timely analysis of information are essential to effective and efficient policing", which means that information gathering is very important for the police. The police have many ways of gathering information; however, this is regulated by the Regulations of Investigatory Powers Act 2000 in the UK.

The main police methods of collecting information are "routine, tasked and volunteered information". Amongst these methods, the tasked and routine methods of data collection are what focus on the importance it plays in keeping children safe. The fear of misleading information and concerns around data protection is minimized by the fact that information that police hold must adjust to data quality principles, which are "accurate, adequate, relevant and timely". Therefore, information and data collected and used by the police for child protection and safeguarding purposes are safe and are used only when required and as specified by the Data Protection Act 2018 [13] with a designated data protection officer to ensure this is not breached. Controversy and differences in views on police use of AI and big data would continue, however, the fact that the number of the police force is reducing compared to the number of daily online crime suggest that the use of AI and big data will continue to prevail. So long as individual rights are respected and priorities are considered in child protection matters, the use of AI and big data are welcome.

References

1. Baidoo L, Zakrison TL, Feldmeth G, Lindau ST, Tung EL (2021) Domestic violence police reporting and resources during the 2020 COVID-19 stay-at-home order in Chicago, Illinois. JAMA Netw Open 4(9):e2122260. https://doi.org/10.1001/jamanetworkopen.2021.22260
2. Beech M (2022) COVID-19 pushes up Internet use 70% and streaming more than 12%, first figures reveal. [Online] Forbes. https://www.forbes.com/sites/markbeech/2020/03/25/covid-19-pushes-up-internet-use-70-streaming-more-than-12-first-figures-reveal/. Accessed 17 April 2022
3. Bollen J, Mao H, Zeng X-J (2011) Twitter mood predicts the stock market. J Comput Sci 2(1):1–8
4. Bouma H et al (2017) Automatic analysis of online image data for law enforcement agencies by concept detection and instance search. In: Proceedings of the SPIE 10441 counterterrorism crime fighting forensics and surveillance technologies
5. Bromfield LM, Holzer PJ (2008) A national approach for child protection: project report. A report to the Community and Disability Services Ministers' Advisory Council (CDSMAC). Australian Institute of Family Studies, Melbourne
6. Cabinet Office (2018) Government security classifications May 2018. Version 1.1

7. Clark L (2012) Google's artificial brain learns to find cat videos. Wired. https://www.wired. com/2012/06/google-x-neural-network/
8. Commission E. The EU data protection reform and big data: factsheet 2016 [https://publicati ons.europa.eu/en/publication-detail/-/publication/51fc3ba6-e601-11e7-9749-01aa75ed71a1. Accessed March 2022
9. Convention on Cybercrime, Budapest 23.XI.2001, CETS 185. http://conventions.coe.int/Tre aty/Commun/QueVoulezVous.asp?NT=185&CL=ENG. Accessed 20 March 2022
10. DHSC (2013) Child protection: information sharing project. GOV.UK
11. Dai M, Xia Y, Han R (2021) The impact of lockdown on police service calls during the COVID-19 pandemic in China. Polic J Policy Pract 15(3):1867–1881. https://doi.org/10.1093/police/paab007
12. Data Ethics (2021) A data democracy comes with individual data control
13. Data Protection Act 2018
14. De Mauro A, Greco M, Grimaldi M (2016) A formal definition of big data based on its essential features. Libr Rev 65(3):122–135. https://doi.org/10.1108/LR-06-2015-0061
15. De Sousa WG, de Melo ERP, Bermejo PHDS, Farias RAS, Gomes AO (2019) How and where is artificial intelligence in the public sector going? Gov Inf Q 36(4):101392. https://doi.org/10.1016/j.giq.2019.07.004
16. Department for Education (2014) Working with foreign authorities: child protection and courts orders. Child protection: working with foreign authorities—GOV.UK. www.gov.uk
17. Drummond B, Christie L (2022) Sharing public sector data. UK Parliament Post. Number 664. POST-PN-0664.pdf (parliament.uk)
18. Dunst C (2020) Western democracy's problem with authority makes it more vulnerable to Covid-19. Quartz, 1 May. https://qz.com/1847421/why-most-western-democracies-cant-con tain-coronavirus/. Accessed 26 March 2022
19. ECtHR 24 June 2004, Von Hannover v. Germany, App.no. 59320/00, §77
20. ECtHR 16 February 2000, Amann v. Switzerland, App.no. 27798/95, §65; ECtHR 4 May 2000, Rotaru v. Romania, App.no. 28341/95, §43
21. Ellison NB, Steinfield C, Lampe C (2011) Connection strategies: social capital implications of Facebook-enabled communication practices. New Media Soc 13(6):873–892
22. Engstrom DF, Ho DE, Sharkey CM, Cuéllar M-F (2020) Government by algorithm: artificial intelligence in federal administrative agencies. SSRN 3551505
23. Europarl.europa.eu. 2022. [online]. https://www.europarl.europa.eu/RegData/etudes/BRIE/2020/659360/EPRS_BRI(2020)659360_EN.pdf. Accessed 18 April 2022
24. Fan M (2019) Camera power: proof, policing, privacy, and audio-visual big data. In: Camera power: proof, policing, privacy, and audio-visual big data. Cambridge University Press, Cambridge, p iii
25. Favaretto M, De Clercq E, Schneble CO, Elger BS (2020) What is your definition of big data? Researchers' understanding of the phenomenon of the decade. PLoS ONE 15(2):e0228987. https://doi.org/10.1371/journal.pone.0228987
26. GOV.UK (2022) Police workforce, England and Wales: 31 March 2021 second edition. [Online] https://www.gov.uk/government/statistics/police-workforce-england-and-wales-31-march-2021/police-workforce-england-and-wales-31-march-2021. Accessed 2 April 2022
27. Grabosky PN (1998) Technology and crime control. Australian Institute of Criminology, Canberra
28. HIS Markit: the internet of things: a movement, not a market. IoT_ebook.pdf (ihs.com)
29. HM Government (2010) Working together to safeguard children: a guide to inter-agency working to safeguard and promote the welfare of children. Department for Children, Schools, and Families, London
30. Handel DA, Wears RL, Nathanson LA, Pines JM (2011) Using information technology to improve the quality and safety of emergency care. Acad Emerg Med 18(6):e45–e51
31. Harcourt B (2005) Against prediction: sentencing, policing and punishing in an actuarial age. SSRN Electron J

32. Hargittai E, Hsieh YP (2013) Digital inequality. In: Dutton WH (ed) Oxford handbook for Internet research. Oxford University Press, Oxford, pp 129–150

33. Hargittai (2010) Is bigger always better? Potential biases of big data derived from social network sites. ANNALS, AAPSS, 659, May 2015

34. Hargittai E, Hsieh Y-I P(2010) Predictors and consequences of differentiated practices on social network sites. Inf Commun Soc 13(4):515–536, https://doi.org/10.1080/13691181003639866

35. Henman P (2020) Improving public services using artificial intelligence: possibilities, pitfalls, governance. Asia Pac J Public Admin 42(4):209–221. https://doi.org/10.1080/23276665.2020.1816188

36. Henman P (2010) Governing electronically. Palgrave

37. Home Office, Department of Health, Department of Education and Science, the Wels Office (1991) Working together under the Children Act 1989: a guide to arrangements for inter-agency cooperation for the protection of children from abuse. Her Majesty's Stationery Office, London

38. Hunt X, Tomlinson M, Sikander S, Skeen S, Marlow M, du Toit S, Eisner M (2020) Artificial intelligence, big data, and mHealth: the frontiers of the prevention of violence against children. Front Artif Intell 3:543305. https://doi.org/10.3389/frai.2020.543305

39. IWF (2020) Annual report 2020—Face the facts. Face the facts I Internet Watch Foundation Annual Report 2020 I IWF

40. Kadar C, Maculan R, Feuerriegel S (2019) Public: decision support for low population density areas: an imbalance-aware hyper-ensemble for spatio-temporal crime prediction. Decis Support Syst 107

41. Kebbell MR, Milne R (1998) Police officers' perceptions of eyewitness performance in forensic investigations. J Soc Psychol 138(3):323–330. https://doi.org/10.1080/00224549809600384

42. Klumpp M (2018) Automation and artificial intelligence in business logistics systems: human reactions and collaboration requirements. Int J Logist Res Appl 21:224–242. https://doi.org/10.1080/13675567.2017.1384451

43. Koops B (2009) Technology and the crime society: rethinking legal protection. Law Innov Technol 1(1):93–124

44. Koops B (2013) Police investigations in Internet open sources: procedural-law issues. Comput Law Secur Rev 29(6):654–665

45. Leveson NG (1995) Safeware: system safety and computers. Addison-Wesley, Boston

46. Mac Manus S (2021) Dialogues about data: building trust and unlocking the value of citizens' health and care data. Nesta

47. Manthorpe J, Stevens M (2015) Adult safeguarding policy and law: a thematic chronology relevant to care homes and hospitals. Soc Policy Soc 14(2):203–216. https://doi.org/10.1017/S1474746414000128

48. Marshall H (2022) Policing county lines: responses to evolving provincial drug markets. By Jack Spicer (Springer, 2021, 253 pp). Br J Criminol 62(2):525–527. https://doi.org/10.1093/bjc/azab124

49. McCahill M, Norris C (2002) CCTV in London. Working paper no. 6, Urbaneye. http://www.urbaneye.net/results/results.htm

50. Medical Academy of Royal Colleges (2019) Artificial intelligence in healthcare. Medical Academy of Royal Colleges, London

51. Medicalxpress.com (2022) Study shows domestic violence reports on the rise as COVID-19 keeps people at home. [Online] https://medicalxpress.com/news/2020-05-domestic-violence-covid-people-home.html. Accessed 26 March 2022

52. Mehrotra D, Mattu S, Gilbertson A, Sankin A (2018) How we determined predictive policing software disproportionately targeted low-income, black, and Latino neighborhoods. A trove of unsecured data allowed the first-ever independent analysis of actual crime predictions across the US by the self-described software leader

53. Mena J (2016) Machine learning forensics for law enforcement, security, and intelligence. Auerbach Publications, New York, NY. https://doi.org/10.1201/b11026

54. Ministry of Justice (2022) Achieving best evidence in criminal proceedings Guidance on interviewing victims and witnesses, and guidance on using special measures

55. Mohler G, Bertozzi AL, Carter J et al (2020) Impact of social distancing during COVID-19 pandemic on crime in Los Angeles and Indianapolis. J Crim Justice 68:101692. https://doi.org/10.1016/j.jcrimjus.2020.101692

56. Munro E (2011) The Munro review of child protection: final report. A child-centred system. Department for Education, London. Retrieved from Munro review of child protection: a child-centred system - GOV.UK (www.gov.uk)

57. NHS Digital (2021) Child protection—information sharing project

58. NSPCC Learning (2022) Statistics on child sexual abuse | NSPCC learning. [Online] https://learning.nspcc.org.uk/research-resources/statistics-briefings/child-sexual-abuse. Accessed 27 March 2022

59. National Audit Office (2019) Challenges in using data across government

60. Norris C, Armstrong G (2020) The maximum surveillance society: the rise of CCTV, 1st edn. Routledge, London. https://doi.org/10.4324/9781003136439

61. ODI (2020) Getting data right: perspectives on the UK National Data Strategy 2020

62. Ons.gov.uk (2022) Children's online behaviour in England and Wales—Office for National Statistics. [Online] https://www.ons.gov.uk/peoplepopulationandcommunity/crimeandjustice/bulletins/childrensonlinebehaviourinenglandandwales/yearendingmarch2020. Accessed 2 April 2022

63. Pace (2014) The relational safeguarding model: best practice in working with families affected by child sexual exploitation

64. Pace (2014) Living with child sexual exploitation: keeping your family together

65. Pace (2014) Emma Palmer and Peter Jenkins, parents as partners in safeguarding children: an evaluation of pace's work in four Lancashire child sexual exploitation teams, October 2010–October 2012

66. Pace (2015) Keeping it together: a parent's guide to coping with CSE

67. Pace (2015) Working with the police: the role of parents in child sexual exploitation investigation. Working-with-the-Police-final.pdf (paceuk.info)

68. Poddar S, Mukherjee U (2020) Ascending child sexual abuse statistics in India during COVID-19 lockdown: a darker reality and alarming mental health concerns. Indian J Psychol Med 42(5):493–494. https://doi.org/10.1177/0253717620951391.PMID:33414605;PMCID:PMC7750843

69. Raaijmakers S (2019) Artificial intelligence for law enforcement: challenges and opportunities. IEEE Secur Priv 17(5):74–77. https://doi.org/10.1109/MSEC.2019.2925649

70. Ramaswamy S, Seshadri S. Children on the brink: risks for child protection, sexual abuse, and related mental health problems in the COVID-19 pandemic. Indian J Psychiatry 62 (Suppl 3):S404–S413. https://doi.org/10.4103/psychiatry.IndianJPsychiatry_1032_20. Epub 2020 Sep 28. PMID: 33227060; PMCID: PMC7659798

71. Ritchie F (2021) Microdata access and privacy: what have we learned over twenty years? JPC 11

72. Robert G (2013) Participatory action research: using experience-based co-design to improve the quality of healthcare services. Understanding and using health experiences–improving patient care, pp 138–150

73. Saif H, Dickinson T, Kastler L, Fernandez M, Alani H (2017) A semantic graph-based approach for radicalisation detection on social media. In: Proceedings of the extended semantic web conference (ESWC 2017), pp 571–587

74. Saunders J, Hunt P, Hollywood JS (2016) Predictions put into practice: a quasi-experimental evaluation of Chicago's predictive policing pilot. J Exp Criminol 12(3):1–25

75. Schwab K, Malleret T (2020) The great reset. In: World economic forum, Geneva, vol 22

76. Spicer J (2021) The policing of cuckooing in 'County Lines' drug dealing: an ethnographic study of an amplification spiral. Br J Criminol 61(5):1390–1406. https://doi.org/10.1093/bjc/azab027

77. Stickle B, Felson M (2020) Crime rates in a pandemic: the largest criminological experiment in history. Am J Crim Justice 45(4):525–536

78. Sun TQ, Medaglia R (2019) Mapping the challenges of artificial intelligence in the public sector. Gov Inf Q 36(2):368–383. https://doi.org/10.1016/j.giq.2018.09.008

79. Support A (2022) Management. [Online] Naht.org.uk. https://naht.org.uk/AdviceSupport/Topics/Management/ArtMID/755/ArticleID/127/Keeping-children-safe-in-education-2020. Accessed 25 April 2022

80. Tarleton A (2009) FactCheck: how many CCTV cameras? How two streets in Putney spawned a nationwide statistic that runs and runs

81. The Council of Europe Commissioner for Human Rights: Recommendation Unboxing Artificial Intelligence: 10 steps to protect Human Rights (May 2019). https://rm.coe.int/unboxing-artificial-intelligence-10-steps-to-protect-human-rights-reco/1680946e64

82. Themarkup.org (2022) Crime prediction software promised to be free of biases. New data shows it perpetuates them—the markup. [Online] https://themarkup.org/prediction-bias/2021/12/02/crime-prediction-software-promised-to-be-free-of-biases-new-data-shows-it-perpetuates-them. Accessed 25 April 2022

83. Tumasjan A, Sprenger TO, Sandner PG, Welpe IM (2010) Predicting elections with Twitter: what 140 characters reveal about political sentiment. Paper presented at the Fourth international AAAI conference on weblogs and social media, Washington, DC

84. Vitorino P, Avila S, Perez M, Rocha A (2018) Leveraging deep neural networks to fight child pornography in the age of social media. J Vis Commun Image Represent 50:303–313

85. Wahl B, Cossy-Gantner A, Germann S, et al (2018) Artificial intelligence (AI) and global health: how can AI contribute to health in resource-poor settings? BMJ Glob Health 3:e000798

86. Wears RL, Leveson NG (2011) "Safeware": safety-critical computing and health care information technology

87. Wolak J, Finkelhor D, Mitchell KJ (2008) Online "predators" and their victims: myths, realities, and implications for prevention and treatment. Am Psychol 63:111–128

88. Wright A, Holliday R (2005) Police officers' perceptions of older eyewitnesses. Leg Criminol Psychol 10(2):211–223

89. Wu X, Zhang X (2016) Automated inference on criminality using face images. http://arxiv.org/abs/1611.04135

Beyond the Surface Web: How Criminals Are Utilising the Internet to Commit Crimes

Kaycee Jacka

Abstract The internet has revolutionised society and plays a key role in users lives. The rapid advances in technology means that now more than ever, users have the right to anonymity online, a service which is greatly invaluable to some, however, is leading to illicit behavior in others. Security threats and the distribution of illegal materials and substances continue to take place online, with a combination of legal loopholes and advancing anonymity subsequently leading to slow investigatory processes. This research considers how criminals are using services to their advantage and remaining largely undetected by the criminal justice system. This research also considers how ethically the criminal justice system works when dealing with its investigations with recommendations for the future. The growing use of cryptocurrencies are also considered with its security advantages for users but how the virtually anonymous features are enticing for criminal activity. This research considers whether digital forensics is keeping up with demand within the criminal justice system and whether new services with the standardisation and collaboration of governments is required to aid further investigations.

Keywords Deep web · The Onion Router · Digital forensics · Bitcoin

1 Introduction

In everyday use of the internet, we use what is referred to as the surface web; however, this only makes up 4% of what the internet has to offer. This gives users access to websites such as Google, YouTube, social media, or any website that can be accessed via a link. "Searching on the Internet today can be compared to dragging a net across the surface of the ocean. While a great deal may be caught in the net, there is still a wealth of information that is deep, and therefore, missed" [4]. Comparably, as

K. Jacka (✉)
Department of Criminology, Sociology and Social Policy, School of Social Sciences, Swansea University, Singleton Park, Swansea SA2 8PP, UK
e-mail: 854956@Swansea.ac.uk
URL: http://www.swansea.ac.uk

specialist equipment is used to explore deeper parts of the ocean, it is used to explore the web. The deeper you explore and seek to find, the more knowledge is needed on how to navigate the 96% of the internet referred to as—the deep web [13]. This gives users access to anything that can't be accessed by a link or password protected domains such as private emails or workplace servers. Making up 1% of the deep web is the dark web, accessible only via specialist tools and equipment that protect the user's identity. Considering the scale and invitation of anonymity that the dark web has to offer, the advantages of using it for purposes such as communication platforms are endless.

However, with advantages come disadvantages and with every legal and well-meaning platform, the dark web has become home to illegal services ranging from child pornography and human trafficking to fraud and data selling. Added complexities have emerged making it easier than ever before to partake in crime online anonymously. The emergence of cryptocurrencies has enabled users to exchange goods in a completely anonymous nature meaning the selling and purchasing of drugs and weapons, human trafficking and money laundering is all possible. There is no need to provide any form of identification to purchase bitcoin and it cannot be traced back to a user making it a great asset to those who advocate for privacy, especially online. Government agencies have had to come up with unique and inventive ideas to track down these online criminals who can be anywhere in the world and usually have the knowledge and technology to distort their IP addresses, not leaving behind traces that other internet users would. These have posed ethical questions as to how far agents can go to stop illegal activity before it reaches entrapment and violates laws. Since this field of criminal activity is still a relatively new one that is rapidly advancing, establishing laws such as the Computer Misuse Act (2003) and following the RIPA Act (2000) has been vital in convicting online offenders. This chapter aims to understand how criminals are utilising all the mentioned resources to take part in illegal activity, and largely getting away with it. I seek to understand the shortfalls and loopholes within the criminal justice system and provide recommendations on how to address these ethically whilst keeping up with rapid technological advances.

2   Background

2.1   The Layout and Purpose of the Web

The concept of a worldwide communications and information platform was initially discussed in memos written by J. C. R Licklider in 1962, whereby the idea of the Galactic Network was envisioned. In essence, what Licklider was laying out is extremely similar to what is known as the internet today. A "globally interconnected set of computers through which everyone could quickly access data and programs from any site" [18]. In 1966, a proposal was put forward for a digital communications network whereby Donald Davies coined the word packet, a "small subpart of the message the user wants to send" and later introduced "the concept of an "interface computer" to sit between the user equipment and the packet network". Though a lot simpler than what is known of the internet today in 2021, these beginnings remain the foundation of the technology that is used to communicate and store data online. By 1969, the 'connecting' began, meaning four computers were then able to communicate with each other via the 'ARPANET' [2]. Through development of the ARPANET, the first mail communications were sent in 1972 via traditional circuit switching methods and "special purpose interconnection arrangements between networks were another possibility". In 2021, users now have access to a range of websites with a multitude of purposes at the click of a finger. Search engines, such as "Ask Jeeves", "Google", "MSN", and "Yahoo!" can take a user to a host of websites such as books, online shopping, social media etc. These are public webpages whereby your I.P address is tracked across sites and does not require authentication or permission to access. "The other content is that of the Deep Web, content that has not been indexed by traditional search engines such as Google." [9]. Using specialist software such as The Onion Router (TOR), users can navigate the deep web to find otherwise hidden webpages. TOR was created with legitimate purposes by the U.S. Naval Research Laboratory as a tool for anonymously sending sensitive information online. "Those who run dark websites that end in ".onion" are able to hide their identities and locations from most, if not all, Internet users" [7] making it extremely difficult once found, to prosecute those that are involved in the selling or distribution of illegal content or materials.

2.2 The Layout and Purpose of the Web

Since the internet is used for communications and intelligence and stores a lot of information, there are ways in which this has been exploited. Cyber trespassing occurs when crossing boundaries into other people's virtual property and/or causing damage to it. Whilst a relatively simple concept, establishing when a boundary has been crossed and whether there was malicious intent behind the action is still a debated one. Cyber theft is another type of crime committed whereby an individual steals online property including money, personal information to be used for fraud, intellectual property, and piracy. According to NFIB Cybercrime Assessment 2020/21 the top three instances of cybercrime were:

- Hacking social media and email—13,948 reports
- Computer Virus/Malware/Spyware—7,794 reports
- Hacking Personal—5,587 reports.

The key enablers of this were 'phishing emails' in which an individual clicks a link to what seems like a credible website, follows the link and enters their sensitive information. The report also found that users had weak or the same passwords across multiple platforms, meaning the hackers were able to access more than initially intended. Contrastingly to stereotypical fraud whereby the high-risk victims were aged 65+ [22], the high-risk age categories identified by the National Fraud Investigations Bureau for cybercrimes were aged between 20 and 39. These crimes typically target users and take place on the surface web. With an added layer of anonymity that is provided by the deep web, far more malicious crimes can take place. The assumed anonymity of using a router such as TOR to distort the users IP address, making them harder to locate means that the deep web has become a gateway for criminal activity. With the advances of digital currencies such as bitcoin, it means that users can trade illicit content such as child pornography as well as use sites for human trafficking and the selling/purchasing of illegal drugs and weapons.

3   Challenges

3.1   Ethical Hacking or Blatant Computer Misuse?

There is a blurred line of ethics when it is discussed what information should be public knowledge and what should remain private within organisations. Simply defined, hacking is "the unauthorised use of, or access into, computers or networks by exploiting identified security vulnerabilities" [6]. When looking at the case of Raphael Gray, the teenager who hacked into and published over 6,500 pieces of stolen credit card information, there is a clear ethical challenge. This was done as a statement to corporations about their weak security defences. Gray considered his actions not only ethical but needed and encouraged others to do as he did and controversially received praise from a lot of the public. His defence included that obtaining this information was legal as the on these sites, "because there was no warning that access was prohibited" [12]. The cost of closing the accounts and redistributing cards to those who had their information stolen by Gray, was set at an estimated cost of $3m [3]. The average cost of a data breach reached U.S. $6.53 million in 2015 [11]. When the Sony PlayStation Network was hacked, "compromising the personal and financial information of more than 77 million user accounts" [21] in 2011, it was the largest recorded data breach of the time with the direct costs estimated at $171 million.

3.2   When Cybertrespass Is a Crime

As previously mentioned in the case of Raphael Gray, establishing that a cybercrime has taken place can be challenging however the introduction of the Computer Misuse Act (2003) has sought to clarify to prevent cases like this happening again. Whilst establishing cyber trespassing, if there is no contract established between the two parties it is difficult to decipher whether a crime has taken place. In a review of Unauthorised access as legal mechanisms of access control, Wong sets out that;

"Like their US federal counterpart, the UK and Singapore computer misuse statutes also reveal a property-based notion of computer crime, as well as a lack of clarity or definition as to the concept of 'unauthorized access'" [24]. This is where the first legal challenge of cybercrimes begins. As there is no geographical crime, it cannot simply be compared to breaking into a house and classifying the act as trespassing, as there is no physical house only virtual boundaries and walls. However, much like the physical counterpart, clues are left behind and it is the role of Digital Forensic analysts to find these traces left behind. Causing a computer to perform a function can be as simple as opening an unauthorised file to as complex as hacking and stealing information.

3.3 Ethical Considerations of Police Using the Dark Web

Honey traps are a technique used by undercover agents to gain access to a user's IP address if they are committing illegal activities online. The ethical use of honey traps have been debated due to the lack of evidence for its efficacy. "There is no defence of entrapment in English law, but it is considered to be an abuse of the process of the court for state agents to lure a person into committing illegal acts and then seek to prosecute him for doing so" [20]. In 2015, a notorious darknet site called 'Playpen', hidden through the TOR network was uncovered by the FBI. "Playpen hosted 215,000 user accounts and distributed 50,000 images of child pornography before the site was taken down" [8]. After its discovery, the FBI infiltrated the site and maintained it for 14 days in an attempt to monitor its users hoping it would lead to arrests where the FBI injected the site with malware to crack TOR's anonymity of IP addresses. "The Justice Department has said that children depicted in such images are harmed each time they are viewed, and once those images leave the government's control, agents have no way to prevent them from being copied and re-copied to other parts of the internet" [14]. According to the United States Government statistics, as a direct result of the operation to seize the darknet site Playpen, there were 350 U.S. arrests and 548 international arrests. When directly compared to the number of users on the site, this means that 0.42% were arrested and of those 200 are active prosecutions as of May 2017.

3.4 Bitcoin and the Illicit Use of Cyber Currencies

The emergence of bitcoin was not only revolutionary in the tech world but changed societies globally. To have use of a currency that isn't controlled by any national government was a first and for liberal thinkers, a step in the right direction for freedom of privacy. The development of a digital currency also revolutionised the use of the dark web, meaning that the layers of anonymity already created were greater facilitated by the means to make untraceable transactions via bitcoin. Bitcoin "has a value proportional to the credibility attributed to it, the fees involved in the transactions are low, have no limits of territorial use, cannot be frozen or confiscated and have no prerequisites for use or limits imposed per transaction (any amount may be transferred to any person, by any person, to any person, without prior authorization or further justification) [15]. This is the greatest invitation for users of bitcoin, from a simple set up with low fees and no risk, it is the safest way to trade online, along with its complete anonymity. The introduction of cyber currencies has allowed online transactions to be treated with the same level of anonymity as cash transactions in the real world, which is welcomed by many that advocate the right to privacy of expenditure. However, bitcoin quickly became the number one currency used on the dark web and has been linked to many online crimes. Most notorious is that of Silk Road and its seizure by the FBI in October 2013. The darknet site was large on a scale not seen before and was an intricate, well organised operation that quickly grew into a community that dealt with the trade of illegal substances. According to the FBI's criminal complaint filed in Ross Ulbright's trial, the Silk Road market had almost 150,000 buyers and almost 4,000 vendors [23]. Since the trading involved the sole use of bitcoin, detecting any venders or buyers was a virtually impossible task due to the complex combination of TOR and bitcoin with users taking great care to conceal their IP addresses. "Silk Road used tumbler to process Bitcoin transactions as well as tracking individual transactions through Bitcoin blockchain. When a buyer buys something and makes the payment on the website, the tumbler would obscure the link between the buyer's and vendor's Bitcoin addresses" [16].

3.5   Digital Forensics and Criminals

Technology has advanced rapidly which means that lay users of the internet have access to numerous software and Cloud storage not previously available. With the sheer number of users of the Cloud, Google Drive and Dropbox a new challenge has emerged in the field of Digital Forensics. Unlike traditional storage, when information is stored on the cloud it is distributed into multiple nodes rather than a single node. "Due to the distributed nature of cloud services, data can potentially reside in multiple legal jurisdictions, leading to investigators relying on local laws and regulations regarding the collection of evidence" [10]. Not only does this cause legal complications in compiling evidence, but it also leads to significant time delays. The chain of custody that must be followed becomes more difficult due to the nature of multiple nodes making traditional DF technology redundant. This creates a real 'needle in a haystack' scenario for investigators and criminals are utilising these advances in technology more and more to conceal files and remain undetected, without the use of the deep web.

4 Recommendations

4.1 Rectifying the Mistakes of the Playpen Operation

Regarding the darknet site Playpen seized by the FBI in 2015, there are several steps which could have been taken to protect the vulnerable children subject to online abuse. Whilst the investigation was effective in successfully identifying or rescuing 55 American children [8], there are clear safeguarding risks that were not considered. By allowing the website to run for a further 14 days, the FBI allowed the victims of the online abuse to have their material downloaded and possibly redistributed an infinite number of times. The priority of the FBI should have been to safeguard these victims and not provide further harm, as they did. Technology is ever advancing, and an operation could have been as equally effective using such advances. The use of deep fakes or CGI rather than real images of abused children could have reduced harm to the victims resulting in a more ethical investigation. There is no significant data that determines honey traps as the most effective method of policing the dark web and until data supports its use then alternatives should be fully explored before its complete implementation as a standard operation procedure of policing.

4.2 The Next Steps of Digital Forensics

Digital forensics is an extremely important field of technology and must continue to adapt itself to advancing software to adequately support the criminal justice system. "Since December 2010, in the Netherlands a new approach is used for processing and investigating the high volume of seized digital material" [1] using 'Digital Forensics as a Service' or 'DFaaS'. Due to the introduction of the storage spaces such as the Cloud, the time it takes for digital forensic analysts to retrieve information can be an extremely lengthy process. Most crimes now include some cyber element in which an analysts will be brought in to aid the police. However, when they are brought in and what they are asked to investigate varies. This model, which works as an extended process to typical digital forensics, claims to save an invaluable amount of time harvesting data for an investigation. "At a large scale it makes sense to implement a central system that can be used by multiple departments. With this model, it is expected that the system spends less time idling and more time processing data. If digital material becomes available sooner in the investigation, it can be used to form hypotheses instead of only using it to test hypotheses" [1]. There is a clear need for and gap in the market for resources such as these and as criminals get more technologically complex. It is time for the justice system to make provisions, even if this means outsourcing services to better aid their digital investigations. The process is by no means perfect however more funding and research in this field is an investment worthwhile to aid investigators and create a smoother process for criminal proceedings.

4.3 The Standardisation of Law for Cryptocurrencies

There is a clear lack of deterrence for online criminals due to the unclarified legality of cryptocurrencies. Although currencies such as bitcoin are converted from 'real' state currencies, they are still not regarded as a form of tender in most countries creating a legal loophole when cases of money laundering using bitcoin arise. As more users adopt cryptocurrencies as forms of transaction, at the very least there should be standardised legal definitions globally to enable criminal justice systems to prosecute those using it illegitimately. There also needs to be a consideration for the global carbon footprint effects that arise due to 'mining'. It is estimated, that if bitcoin alone were a country, it would use more electricity that Argentina [5]. Bitcoin is arguably one of the most stable currencies to protect against fraud and can be utilised by users safely and legitimately. The mining process verifies each transaction which "makes it extremely difficult for bitcoins to be double-spent or counterfeited" [17] this is a feature that credit card companies currently lack.

5 Discussion

There are clearly many different resources currently available to users to facilitate cybercrimes and new technology is advancing rapidly. The field of Digital Forensics is becoming increasingly complicated due to a number of factors, and it is becoming easier for cyber criminals to conceal their data. "Evidential data is no longer restricted to a single host but instead spread between different or virtual locations, including: online social networks, cloud resources, and personal network–attached storage devices" [19]. New advances, including DFaaS should be a welcomed collaboration to combat the growing number of cyber facilitated crimes to ensure criminal justice services are able to keep up the demand and ultimately reduce harm to users creating a safer space. The investigation process has clear flaws when identifying illicit websites, especially on the dark web. A standardised way of proceedings is necessary, not only to mitigate victim harm but again, to aid investigators when these websites are discovered or reported. Cyber law is also a field that requires more collaboration. Since it is difficult to establish the geographical elements of cybercrime, it is in the interest of all governments to ensure that they are working with the same or similar proceedings due to the legality loopholes that offenders seem to uncover. There is a clear need for the deep web and the services it provides are invaluable when used legally. As more advocates for online privacy arise, it is likely that an increasing number of users will migrate to using services such as TOR or I2P for everyday use. When the number of users grow and begin to protect their anonymity online, digital investigators will have to develop new means in which they are able to detect users they believe are exploiting the anonymity features. As cryptocurrencies are concerned, they are still a relatively new concept to many users. With the many protective features of bitcoin, its stability is dependent on its users, and it is uncertain at this time whether it will ever be able to function as an everyday global currency. The collaboration of governments will have an impact on cryptocurrencies, with many observing El Salvador as a model of its success or failure.

6 Conclusion

The deep web's sole purpose is clearly not to aid criminal activity, and it has many practicalities for its users and as an invaluable business resource for both corporations and governments. It is clear however, that many of its features are designed in a way that easily facilitates illicit activity. There must be more clarity concerning cyber laws and computer misuse for all users to determine illicit activity quickly and effectively. As anonymity in all service arise, digital forensics must adapt its method of investigations to limit criminal behavior and safeguard all users of the web. Further monitoring of cryptocurrencies is needed to detect trends in who is using them and the extent to which it is being used for legal or illegal services. Digital investigators must continue to monitor the dark web as long as illicit materials are being shared, carefully considering not to cross boundaries and document their investigations clearly and thoroughly allowing users to still maintain their rights to anonymity online.

References

1. Banda R, Phiri J, Nyirenda M, Kabemba M (2019) Technological paradox of hackers begetting hackers: a case of ethical and unethical hackers and their subtle tools. Zambia ICT J 3(1):40–51. https://doi.org/10.33260/zictjournal.v3i1.74
2. Bay M (2019) Conversation with a pioneer: Larry Roberts on how he led the design and construction of the ARPANET. Internet Hist 3(1):68–80. https://doi.org/10.1080/24701475.2018.1544727
3. BBC News I WALES I Teen hacker escapes jail sentence (2001). http://news.bbc.co.uk/1/hi/wales/1424937.stm. Accessed 3 Jan 2022
4. Bergman M (2001) White Paper: the deep web: surfacing hidden value. J Electron Publ 7(1). https://doi.org/10.3998/3336451.0007.104
5. Cambridge Bitcoin Electricity Consumption Index (CBECI) (2021). https://ccaf.io/cbeci/index. Accessed 10 Jan 2022
6. Cybercrime—prosecution guidance I The Crown Prosecution Service (2021). https://www.cps.gov.uk/legal-guidance/cybercrime-prosecution-guidance. Accessed 3 Jan 2022
7. Dingledine R, Mathewson N, Syverson P (2004) Tor: the second-generation onion router. ACM Digital Library
8. Farivar C (2017) Creator of infamous Playpen website sentenced to 30 years in prison. https://arstechnica.com/tech-policy/2017/05/creator-of-infamous-playpen-website-sentenced-to-30-years-in-prison/. Accessed 11 Jan 2022
9. Finklea K (2017) Dark web. Congressional Research Service. http://www.crs.gov
10. Focus F (2016) Current challenges in digital forensics—forensic focus. https://www.forensicfocus.com/articles/current-challenges-in-digital-forensics/. Accessed 11 Jan 2022
11. Goode S, Hoehle H, Venkatesh V, Brown S (2017) User compensation as a data breach recovery action: an investigation of the Sony PlayStation network breach. MIS Q 41(3):703–727. https://doi.org/10.25300/misq/2017/41.3.03

12. Gray (2001) https://www.theguardian.com/technology/2001/jul/06/security.internetcrime
13. Hatta M (2020) Deep web, dark web, dark net. Ann Bus Adm Sci 19(6):277–292. https://doi.org/10.7880/abas.0200908a
14. Heath B (2016). https://eu.usatoday.com/story/news/2016/01/21/fbi-ran-website-sharing-thousands-child-porn-images/79108346/. Accessed 11 Jan 2022
15. How do bitcoin and crypto work? | Get started with Bitcoin.com (2016). https://www.bitcoin.com/get-started. Accessed 6 Jan 2022
16. Kethineni S, Cao Y, Dodge C (2017) Use of bitcoin in darknet markets: examining facilitative factors on bitcoin-related crimes. Am J Crim Justice 43(2):141–157. https://doi.org/10.1007/s12103-017-9394-6
17. Lee D (2015) Handbook of digital currency. Bitcoin, innovation, financial instruments, and big data. Elsevier Inc., London
18. Leiner B, Cerf V, Clark D, Kahn R, Kleinrock L, Lynch D et al (2009) A brief history of the internet. ACM SIGCOMM Comput Commun Rev 39(5):22–31. https://doi.org/10.1145/1629607.1629613
19. Montasari R, Hill R (2019) Next-generation digital forensics: challenges and future paradigms. In: 2019 IEEE 12th international conference on global security, safety and sustainability (ICGS3). https://doi.org/10.1109/icgs3.2019.8688020
20. Nexis L (2021) Entrapment | legal guidance | LexisNexis. https://www.lexisnexis.co.uk/legal/guidance/entrapment. Accessed 11 Jan 2022
21. Richmond S, Williams C (2011) Millions of Internet users hit by massive playstation data theft | alternative | before it's news. https://beforeitsnews.com/alternative/2011/04/millions-of-internet-users-hit-by-massive-playstation-data-theft-591301.html. Accessed 11 Jan 2022
22. UK A (2021). https://www.ageuk.org.uk/latest-press/articles/2019/july/older-person-becomes-fraud-victim-every-40-seconds/. Accessed 10 Jan 2022
23. USA v. Ross Ulbricht (Southern District of New York 2013)
24. Wong M (2006) Cyber-trespass and "unauthorized access" as legal mechanisms of access control: lessons from the US experience. Int J Law Inf Technol 15(1):90–128. https://doi.org/10.1093/ijlit/eal014

The Role of the Internet in Radicalisation to Violent Extremism

Olivia Bamsey and Reza Montasari

Abstract This chapter critically examines the role that the Internet and the Internet of Things (IoT) play in violent extremism. The chapter specifically focuses on arguments surrounding radicalisation as a pathway to terrorism and how individuals become radicalised due to different radicalisation processes and theories. Based on this critical analysis, the chapter argues that the Internet plays a key role in radicalisation to violent extremism due to several approaches. Issues highlighted in this chapter mainly focusses on the struggles of social media (SM) moderation and regulations. Furthermore, the chapter discusses how the IoT can aid SM regulation to reduce online radicalisation to violent extremism through artificial intelligence (AI) and machine learning (ML) technology. It will be argued how AI and ML could be deployed to reduce terrorist content online, making social media platforms (SMPs) a safer cyber space for individuals to operate within.

Keywords The Internet of Things · Cyber terrorism · Radicalisation · Violent extremism · Social media platforms · Artificial intelligence · Machine learning

1 Introduction

The Internet plays a significant role in everyday life, which allows a vast majority of the population to be connected instantaneously, regardless of geographical location [79]. The Internet has risen exponentially since the 1990s, and today supplies terrorists and extremists the capabilities of committing serious threat and attacks on the

O. Bamsey (✉)
New York, USA
e-mail: oliviabethbamsey@gmail.com

R. Montasari
Department of Criminology, Sociology and Social Policy, School of Social Sciences, Swansea University, Swansea, Wales, UK
e-mail: Reza.Montasari@Swansea.ac.uk
URL: http://www.swansea.ac.uk

surrounding population which was not available to them pre-1990 [71]. Weimann [92] states there are six ways how the Internet facilitates violent extremism. One way in which the Internet is used in violent extremism is radicalisation. To discuss the concept of radicalisation more effectively, there must be a clear definition. However, currently, a universal definition does not exist, making it difficult to enforce international agreements when tackling radicalisation and extremism [35]. Despite the absence of a universal, agreed upon definition, radicalisation can be described as a process in which a person adopts extremist views and progresses towards engaging in violent behaviours [45]. This is in line with The Prevent Strategy policy's [50, p. 108] definition of radicalisation. According to The Prevent Strategy [50], radicalisation is the "process by which a person comes to support terrorism and forms to extremism leading to terrorism". The remaining ways in which the Internet can be used to ease extremism [92] will be discussed later in this chapter.

Whilst there are various forms of extremism such as domestic (campaigning for animal rights) and non-violent extremism, this chapter will focus only on violent extremism. The British Government define extremism as a "vocal or active opposition to fundamental British values, including democracy, the rule of law, tolerance of different faiths and beliefs, individual liberty and mutual respect" [50, p. 107]. Extremists can be characterised as political actors who ignore the rule of the law and reject diversity within societies [72]. Striegher [80, p. 79] states that the Crown Prosecution Service (CPS) defines violent extremism as the "demonstration of unacceptable behaviour by using any means or medium to express views which foment, justify or glorify terrorist violence in furtherance of particular beliefs". This is including those who engage in terrorist or criminal violence based on ideological, political, or religious beliefs and foster hatred that leads to violence. Violent extremist behaviours often lead to the act of terrorism [86], another term which fails to be universally defined, but cannot be ignored when discussing violent extremism.

Most experts and researchers agree on the definition given in the Terrorism Act 2000 [83] as an "action that...causes serious violence to a person/people; causes serious damage to property; or seriously interferes or disrupts an electronic system". The terrorist's threat is to influence the government or inject fear into the public society, with an aim to change a political, religious, or ideological belief [50, p. 108]. This chapter will be examining the role of the Internet in radicalisation to violent extremism and will discuss the radicalisation processes individuals experience. It will also address how it is possible for an individual to become radicalised into violent extremism entirely through the Internet without offline communication and physical contact in the outside world. Furthermore, AI and ML technologies will be discussed, specifically how they can be used to decrease extremist material online, which in turn, should decrease the number of individuals who become radicalised through social media.

The remainder of this chapter is structured as follows. Section 2 describes terrorism as a process. Section 3 discusses phases of radicalisation, while Sect. 4 examines phases of online terrorism. Sections 5 and 6 critically investigate the roles that SM and the IoT play in enabling terrorism, respectively. Section 7 discusses some of the benefits of the IoT technology in moderating cyberterrorism. Finally, the chapter is concluded in Sect. 8.

2 Terrorism as a Process

Along with the definition previously mentioned, radicalisation can also be explained as a gradual process, whereby individuals are radicalised by friends offline through the internet, or if they have direct communication with extremist groups [20]. How many steps are in this process, however, differs between various researchers [37, 69, 74]. Sageman's "Bunch of Guys" theory [69] is a group based social psychological process theory of extremism which includes four steps that a person must encounter to become radicalised: a sense of moral outrage; developing a specific worldview; resonating that worldview with personal experiences; and mobilising through inter-active networks. The Internet could be involved in all steps towards radicalisation. A moral outrage and the specific worldview could be derived from information seen online in current news articles which encourages the reader to form an opinion based on what they have seen. It could be related to a past personal experience that is sensi-tive to the individual which influences them to engage further to gain understanding. The Internet would play a role in the final step of Sageman's radicalisation process as individuals mobilise through interactive networks online within chatrooms and social media platforms. The Internet is very accessible, where individuals can communicate with those who obtain similar views to them which could contribute to their radical-isation process [26, 53]. This is not as easy in the offline world as social groups are limited to external environments such as location of the individual's home. However, social groups created online can lead to social events in the offline world. Groups sharing the same ideology online can meet in large numbers which could potentially lead to protest marches in cities, possibly causing disturbances.

Gill [37] also describes radicalisation as a pathway to terrorism and extremist behaviours in four steps as represented in Fig. 1. These include: exposure to propaganda; the experience of a catalyst event; pre-existing social ties which aid recruitment; and in-group radicalisation.

Propaganda is defined as a coordinated attempt to influence others to actively spread a point of view with the aim to change society's views [75]. Propaganda can be viewed online through social media and in many forms of multi-media such as online magazines, videos, and images [94]. Social ties could come from current social groups the individual associates themselves with on social media [54]. In-group radicalisation could be achieved through online communities where individuals share common beliefs; again, this is less likely to occur in the offline world due to external environments [85], but is not limited to meeting with social media friends in the "real" world. Gill allows exposure from external sources to be present, while Sageman [69] implies a sense of moral outrage has come from a personal opinion. Although Sageman's [69] sense of moral outrage may be derived from viewing a news article

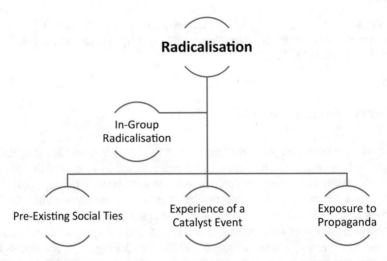

Fig. 1 Steps of radicalisation to terrorism and violent extremism based on [37] description

or video, Gill [37] specifies radicalisation to occur via exposure to propaganda. Comparing the two theories further, Sageman [69] presents the theory in a more internal method compared to Gill [37].

Gill's [37] mention of experience of a catalyst event, past social ties, and in-group radicalisation all deter to external environmental factors compared to Sageman's [69] worldview, resonating it with first hand experiences and mobilising through networks which imply more internal views. This could be reflected in the individuals who are radicalised via Sageman's [69] theory moving to online radicalisation and completing further research via SM, and those who are radicalised via Gill's [37] theory being more extroverted and seeking socialisation from the outside world. This could suggest that those radicalised through the theory of Sageman [69] are more likely to be a lone actor in terrorist activities, and Gill's [37] more likely to act within a group. A theological psychological approach to the radicalisation process shared by Sageman [70] and Wiktorowicz [97] explained how radical religious beliefs are stimulated by group dynamics and lead individuals to become extremists [59]. These beliefs are derived from the echo chamber theory which explains individuals agree with the norms of a group. Individuals decide to listen to opinions they agree with which create the group dynamics [70]. The Internet acts as an echo chamber on SMPs as following a certain group instead of multiple groups can feed certain information [68]. This perspective influenced Silber and Bhatt [74], senior intelligence analysts at New York Police Department, to compile the 'NYPD Model', describing four stages of radicalisation: pre-radicalisation; self-identification; indoctrination; and Jihadization. The Internet could be used and may be a common factor in all steps of this NYPD model into radicalisation through propaganda, social media, videos online, and chatrooms.

3 Phases of Radicalisation

Based on the existing studies [31], radicalised individuals follow three phases during the radicalisation process, including: sensitivity phase, group membership phase and action phase. The sensitivity phase being the time when they engage in radical beliefs and the group membership phase where individuals communicate with those who share common beliefs. The action phase is where the radicalised individual takes part in radical behaviours and act on their beliefs. Alike the other suggested radicalisation processes, these steps can occur online in the same ways [89]. Whether these phases will be reached by the individual is dependent on three levels within the sensitivity phase: micro (individual); meso (group); and macro (societal) level [31]. The micro level is where the individual seeks for information when they feel a loss of identity and belonging [10]. This is likely to occur in the online world as these feelings of significance can be restored by groups such as ISIS by giving new recruits respect and a sense of belonging within the group [31]. It could be suggested that those radicalised through Sageman's [69] theory are more likely to experience this stage due to seeking for significance on SM. This is known as the social identity theory where they have been unable to seek for their identity elsewhere on the outside world [82]. At this stage, extremist groups such as ISIS motivate recruits to strengthen their identity with the group, enabling them to adopt norms and values [49]. As these extremist groups present a strong structure [52], someone who experiences personal uncertainty would be attracted to their ideology [48].

The meso level consists of the external environment the radicalised individual would be surrounded by such as friends, family, and those active in extremist groups [31]. Research shows it is more like for individuals to become radicalised if peers such as friends and family are taking part in learning ideologies too [67]. Gill's [37] theory of radicalisation suggests this, as pre-existing social ties could become a factor of radicalisation. A factor within this level is known as fraternal relative deprivation, where people believe their group has been treated negatively compared to other groups [25]. In context, radical right groups such as the British National Party (BNP), the English Defence League (EDL) and Britain First believe national citizens are treated worse than immigrants [32] and therefore want to highlight and make a change within society where they would perceive they are treated equally or even above them. In the past on SM, the BNP made themselves very public with their views, such as their belief that immigrants were taking 'white British' jobs [41]. These views could have been spread in the offline world, however, would not have gained as much publicity. Unless the party travelled throughout the UK to promote their views, they may not have spread further than a county or city. This is how SM plays such a substantial impact on group policies and beliefs being spread.

The macro level sees the radicalisation process being influenced by societal factors [64]. The macro level can be described by the effect of globalisation and modernisation [24] along with issues of foreign policy. Due to these factors, it is considered that globalisation initiates terrorism and violent extremism [63], and even more so using the Internet. For example, with technology live streaming the 9/11 attacks conducted by al Qaeda, the terrorist organisation received worldwide recognition from the media [43]. Along with the Christchurch shootings being livestreamed on Facebook, this strongly shows how SM needs to become better equipped in regulating terrorist material. Micro, meso, and macro factors have a major influence on whether individuals will continue to become radicalised, and all interconnect with each other. Once these stages have been experienced by the radicalised individual, the group membership and action phase can take place [31].

Alternative views suggest an individual can foster radicalised beliefs either by having adopted few radical ideas themselves or not having extremist views at all at the beginning of the radicalisation process [65]. Information they view online can influence them to undertake extremist views and the Internet is a tool that supplies greater opportunity for planning attacks and violent radicalisation [38]. It must be noted that not everyone who agrees with radical views engage in the physical acts of committing a terror attack [12]. When not actively taking part in these events, the radicalised individuals would take on another key role in the organisation of a terror attack such as motivators at events, propaganda distributors which encourage more individuals to join, or run social media to promote the organisation [9]. Although there is no psychological profile that matches all extremists, research shows those who face socioeconomic disadvantage, government oppression, and mental health are more likely to engage in extremist behaviours [18, 51].

Radical groups share common beliefs [17] such as believing in serious issues within society and have an urge to change these. The issues can be explained through an overlap of perspectives that draw upon psychological, social, political, and economic factors. Political issues such as government oppression can be the root of an extremist ideology [96]. Anger can stem from institutions not dealing with grievances, resulting in engaging in violent behaviours [61]. As aforementioned, a radical group may arrange march protests via online groups which gains members and influences their beliefs onto others. This is known as deindividuation where individuals behave differently in groups and do not view themselves as individuals, they adopt the opinions of the groups as opposed to developing ideas themselves [28]. This also agrees with Gill's [37] theory as opinions are shared with an attempt to radicalise individuals further. Individuals would engage in impulsive, deviant, and violent acts where they believe they cannot be personally identified [33], in this instance, participating in violent extremist behaviours such as protests. Details about marches can be posted online and spread to thousands of users for free and at a quick rate [8]. Without the Internet, distribution of details would occur through word-of-mouth and physical propaganda such as posters, newspapers, and TV coverage [55]. The Internet makes the radicalisation process much more cost-effective and efficient to radical groups by enabling them with a global reach of audience [88]. It must be stated that researchers found more Internet research into extremist behaviour, resulted in more physical con with fellow extremists such as attending marches [38].

4 Phases of Online Terrorism

There are three phases of online terrorism which contributes to violent extremism—the early years (1990, 2006), 'web 2.0' and the regulatory fight era. Tim Berners-Lee invented the World Wide Web in 1989 which led to 16 million users, 0.4% of the world's population, by 1995 [78]. Recent figures by Statista [78] show active online users have increased to 4.66 billion, covering 59% of the world's population. Within the first phase, terrorists were early adopters of the Internet [87]. They were attracted to cyber space as it is inexpensive and developed an increased reach of audience [21]. The 1990s saw the rise of the Radical Right, extremist groups with policies leaning towards conservatism, nationalism, and anti-immigration. This timeframe also matched the rise of the Internet for the Radical Right group, the British Nationalist Party (BNP). They used the Internet to their advantage to gain their following on Facebook. The second phase (2007–2015) saw the introduction of social media and mobile technology, which coincides with the rise in ISIS [66]. Platforms such as YouTube and Twitter saw the first of extremist activity, with around 22,000 Twitter users contributing to support or propaganda distribution for ISIS [11]. Onwards from 2016 saw the beginning of phase three where platforms took an active approach in removing extremist content and groups on social media, which led to ISIS degrading. Twitter suspended more than 200,000 extremist accounts in 2016, leading to the rise of extremist content appearing on end-to-end encrypted messaging platforms such as Telegram [29]. This era also saw the devolvement of the BNP, EDL, and Britain First, as they became permanently banned from Facebook in 2019 due to falling under their "dangerous individuals and organisations" policy [34, 91]. Internet has evolved extremely quickly over the last decade, providing extremist groups with the facilities and materials they require to commit to their success.

According to Weimann [92], there are six ways through which the Internet can be exploited to facilitate violent extremism. These include recruitment, socialisation, communication, networking, mobilisation and coordination as shown in Fig. 2.

Recruitment is successful to extremist organisations by using social media through the Internet [4]. Those seeking for new members to join their group exploit existing grievances in vulnerable users [77]. They reach out to individuals suggesting they can supply a sense of belonging and a positive life within extremist groups such as the ISIS community. This would be attractive for users who believe they lack a sense of belonging in their community [27]. It can be said that many people have been radicalised and encouraged by propagandists who are overseas, therefore would have had to occur online [50]. It is evident that this would not have occurred if the radicalised individual did not view propaganda online or did not have Internet access at all. However, the Home Office [50, p. 13] states evidence suggests some individuals who have been radicalised in the UK had taken part in extremist organisations in the past. Therefore, it may be easier for propagandists to entice those with a history of extremist behaviours to join their organisation. Weimann [92] expresses socialisation, the process of internalising the norms of a group, is also used through the Internet

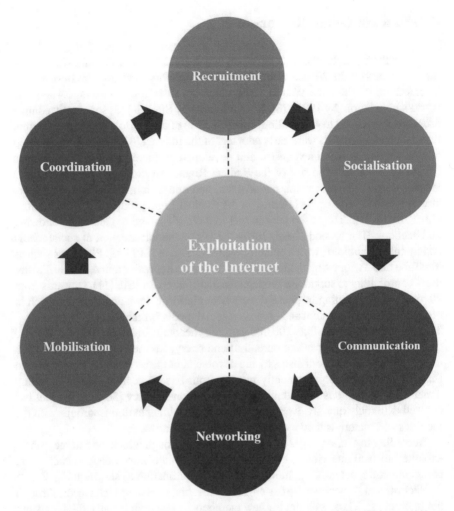

Fig. 2 Six phases through which the Internet can be exploited, as suggested by Weimann [92]

by violent extremists. Through the process of socialisation, Weimann [92] reveals users learn the language of the culture, their roles, and responsibilities in life and what is expected from them. It is possible for group polarisation to occur during the socialisation process. Group polarisation is referred to a group making decisions less rationally, and therefore more extreme than initial ideas of individuals within the group [76]. Users may develop a strong dedication to the online community and withdraw from offline peers if they are socially isolated as they now feel they are a part of a new culture. Socialisation can occur solely online, as the Internet is a safe space for those normally isolated from one another and online interactions fulfil a sense of community [56]. In external environments, it is difficult to find a large group of people who hold the same beliefs in this context. As the Internet provides individuals with instant messaging and reach to all over the world, it gives them the opportunity to find belonging in groups alike.

Weimann also discusses communication as a vital component of terrorist activity online [92]. Communication is mainly heightened within extremist groups in social media to circulate propaganda, recruiting members, raising funds, attempting to normalise extremist views, advising members on how to support the group, and gaining publicity [7]. Organised groups use smaller alt-tech platforms to coordinate mainstream campaigns [95]. Examples of these alt-tech platforms are Gab, and Telegram that impose less strict content moderation rules. Gab is a social media platform known for its far right userbase and free speech capability [30]. Telegram, on the other hand, is an instant messaging service where messages are heavily encrypted and can be deleted as soon as it meets the recipient [5]. This decreases the chance of the message getting intercepted by an unintended user to protect the information being sent. Also links to recruit individuals have an expiration time [15], encouraging potential recruits to join quickly. It is also a safety feature for the organisation in case the link is shared but not used, with an expiration time, it decreases the chance of the group getting caught by authorities. Bloom and colleagues [15, p. 4] state there are three types of Jihadist Telegram users: those who search for content; committed sympathisers and propagandists who are actively creating groups; and official propagandists who create proxies. ISIS use Telegram to participate in chatrooms which are used to recruit new members privately [58].

Two-way interactivity is present within radicalisation as audiences become active participants when they engage and respond to comments and chatrooms online [3]. Multiple researchers highlight the advantages of these alt-tech platforms to extremists in the aim of radicalisation. Advantages to their success include free communication, the end-to-end encryption for heightened security, and instant service of distributing material to a targeted audience [42, 58, 93]. Engaging in networking is another way in which extremists utilise the Internet [92]. Weimann [92] explains how the Internet provides violent extremist groups with instant connection around the world regardless of geographical location and a sense of readiness 24/7 due to push notifications. Global networks are created within large platforms due to their reach. Krona [58] states the Internet gives the capability to extremist groups to operate as a more decentralised organisation as communication can be conducted over networks, this brings Weimann's [92] remarks up-to-date. Social networks are not only used for the exchange of ideologies, but these anonymous platforms share instruction manuals on how to make bombs, poison, and conduct attacks [92]. Active users are encouraged to refer to these documents and act on them in the offline world. These 'how to' guides are too easily accessible within platforms such as Gab. With the addition of users viewing videos of 'successful' attacks due to homemade bombs, it could possibly be even more tempting for individuals to act on this to see if they can achieve what they see in the video, leading to becoming radicalised. Young men can become easily attracted to the visual imagery used in propaganda videos from ISIS, due to the real-life likeness of video games [1, 2].

The propaganda videos consist of very high-quality cinematography used in propaganda videos, like a real-life video game. This kind of platform may be appealing to those who are searching for their identity, faith, or sense of belonging [10]. 'Successful' organised attacks are also portrayed through online magazines developed by terrorist groups. Online magazines are persuasive towards young men as they explain if women and children can provide support for their country, men should participate too. The ISIS magazine 'Dabiq' convinces the reader to engage in acts of violent extremism and persuades individuals to travel to the Middle East. If they cannot commit to this, they are encouraged to perform lone-wolf attacks in their home country [13]. The al Qaeda magazine 'Inspire' is less informed and targeted towards less intellectual individuals which includes instruction manuals and drives readers to act. The fifth, and sixth way Weimann states terrorists use the Internet are known as mobilisation and coordination [92]. The Internet can be used to mobilise followers to become more involved in active roles to assist terrorist activity. Online communication also plays a role in this as it can enable extremist groups to coordinate members to undertake action. This action may be taking part in demonstrations, rallies or engaging in violent extremist behaviour. Coordination of groups on smaller platforms can organise live streams to take place on mainstream platforms [22]. Wyman's six ways the Internet facilitates violent extremism are all intertwined and contribute together to the extremist behaviours they result in. In addition to Wyman's six factors, the use of social media to extremist's aim of radicalisation in the mainstream world such as Facebook and Twitter must be mentioned.

5   Social Media Platforms and Terrorism

Extremists use mainstream SM to increase their reach even further to the wider population [72] because it is inexpensive, easily accessible, and multi-media options are available such as video and image usage. The extremist organisations conduct the same acts on these platforms as they do with Gab—spread ideology, create fear within societies, motivate problems, recruit new members, display propaganda, and provide an ingroup/outgroup narrative. SMPs cannot do anything directly harmful to individuals or societies but can pose a threat on the outside world. The use of images being uploaded online is popular within extremist groups. Photographs are an easier tool for communication which gains quicker attention from the reader compared to words and creates social knowledge [46]. Viewers are also more likely to remember images and the meaning behind them better than text as there are no language barriers [57] and gain a quicker and more positive emotional reaction [40]. Images are likely to be submerged into popular areas of SMPs to gain the maximum attraction. Extremist groups attach hashtags onto images to divert them into trending material online. For example, "#worldcup" was used for propaganda material to appear in the thread of trending tweets in the aim to recruit individuals into extremist groups [62]. The more views extremists gain on their online material, their chance of radicalising individuals increase.

However, it could be believed that SMPs could be ISIS' downfall. As profiles remain high in anonymity, those who claim they will act on extremist ideology may be hiding behind a keyboard and never act upon it. This explains the online disinhibition effect by Suler [81], who states it is easier to speak with no disinhibition online, as communication is delivered differently in the offline world [14]. Therefore, ISIS may believe more people are committing acts compared to reality. Another downfall of social media for ISIS is the possibility that novice users may provide insight to counter terrorism due to their lack of knowledge within these platforms. Inquisitive new users may not be as reluctant as 'experts' within SMPs, which could lead them to accessing links taking them to an unknown source. There is a possibility that external links may be tracked by authorities which could be costly to extremist groups, nevertheless, would be positive for the wider society.

Problems that arise with regulating SMPs include balancing censorship with freedom of speech. It must be remembered that individuals have the right to freedom of expression, however the rules on social media platforms such as Facebook's Community Standards [34] state they must not include hate speech. There is also a lack of clear universal definition of terms such as extremism and terrorism across the globe. This can introduce a conflict of requirements within privacy and security when regulating social media [50]. Due to the increased reach available to extremist groups online, moderators struggle with protecting individuals within the wider community as they should not be seeing extremist content online. Terrorists and extremist groups are actively seeking new ways to take advantage of the Internet and it seems moderators and governments are not as quick to access new tools. Thomas [84, p. 114] also states how "governments cannot control the Internet to the same degree they could control newspapers and TV". There is also a problem within mainstream and alt-tech platforms. Take YouTube for example, their algorithms of suggested videos could function as the pipeline to radicalisation due to videos being presented in front of them even when they do not go actively seeking for it. Individuals should report these videos when prompted with them, so they can be removed to decrease the chance of radicalisation.

6 Social Media Platforms and Terrorism

The IoT consists of devices connected to the Internet which consist of sensors, software, and the ability to transfer data over a network without requiring human interaction [39]. Real-world examples of IoT devices include wearable smart watches, smart TVs, voice assistants such as Amazon Alexa, and conless payments [6]. More recently in technological developments, Artificial Intelligence of Things (AIoT) has become more widely used to achieve more efficient IoT operations [19]. Christensen [19] suggests the IoT manages the devices connected to the Internet, while AI makes the device learn future tasks based on data and experience. AI can be advantageous in developing systems within the health setting, marketing, and social media moderating. However, AI can be used for malicious purposes within the process of radicalisation towards extremism. Although it is expected that AI is extensively used to promote propaganda and spread extremist material online, the assumption is widely understudied and requires more empirical research. This is expected in the future due to the emergence wireless technology and the IoT becoming more pervasive [47]. Evidence suggesting the IoT does not encourage radicalisation derives from Schroeter [73]. The author suggests not enough data can be gathered on SM to create an algorithm which proves an individual has been radicalised online and committed terrorist acts. However, arguing against this, it has been shown that AI has taken place to distribute "Fake News". This was conducted by creating realistic photographs and new accounts to distribute information which avoided detection from social media software which seeks false accounts [90]. Although this was not related to terrorist material, it poses the threat that terrorist and extremist groups can do this in the future to avoid identity detection. AI is also used in the prevention of terrorist material being posted online via social media moderating. Key words can be detected in posts which automatically remove the sensitive material. However, terrorist groups attempt to avoid the detection by inserting punctuation in-between words, and by creating new accounts. However, AI can also falsely remove posts if the material is incorrectly detected as offensive, this remains a fault in the AI system which is expected to be corrected as technological advancements continue to develop.

7 The Internet of Things Moderating Online Terrorism

Countering online extremism has increased dramatically over the past years due to technology evolving and enabling moderators to remove content and protect the wider public. Decreasing the amount of extremist content online requires coordinated response across governments, private companies, and independent regulators [44]. The Internet of Things plays a role aiming to stop extremist activity being posted online on social media platforms. AI and ML technology are two ways that can be used to detect and remove activity without human interaction [60]. AI and ML "learn rules from data, adapt to changes, and improve performance with experience" [16, p. 1]; whereas a content moderator is required to manually remove content. While technological and societal interventions are available and somewhat effective, these interferences alone are unlikely to eliminate terrorism entirely. Also, a study has found 4.8% of detected offensive tweets were misclassified; they contained offensive language which did not involve hateful words [36]. Although this was a small percentage of wrongful detection, the remaining 95.6% was accurately detected and therefore successfully removed. Since 2006, the British government announced its public strategy to counter international terrorism [50]. Their aim was to tackle terrorist use of the Internet as platforms are used to display many radical views, which can influence vulnerable individuals. One example of a team of human moderators who aim to remove online terrorist material are the Counter Terrorism Internet Referral Unit (CTIRU) in the UK. Established in 2010, they challenge the increase of extremist and terrorist related material posted online by removing or modifying content [23]. Since 2010, they have successfully removed over 304,000 items of unlawful internet content and continue to identify those liable for posting harmful material. The unit actively scans the Internet for extremist content as well as researching into reported websites by the public. Automatic content removal using AI and ML algorithms are popular interventions for reducing terrorist related material on social media as it can be removed quickly and reduces the number of users viewing extremist material. However, sometimes it is not removed quick enough. If content is removed success-fully and efficiently, this safeguards human moderators from viewing harmful mate-rial as it could be psychologically damaging. The difficulty in automatically removing posts from word detection is the range of languages the messages can be displayed in. Social media companies would have to recruit bilingual speakers to detect hate speech and extremist posts in different languages. Also, experienced users can avoid word detection not using specific terms that are usually recognised and removed. This makes it difficult for automatic detection to be efficient and therefore content is accessible for longer allowing more people to view the material. Users also post multi-media such as videos, images, and memes to avoid getting detected. Also, if material is posted on the Dark Web, it can be impossible to remove due to the decentralised server. There are other ways online extremism can be countered such as de-platforming, societal and individual interventions.

8 The Internet of Things Moderating Online Terrorism

It can be stated that the Internet can play a sole role in radicalising individuals into violent extremism as offline external factors are not always required; however, they can lead to social events in the offline world such as protest marches through the communication of online advertisement. The role of the Internet most definitely affects radicalisation in individuals into violent extremism described through the explanation from Weimann [92]. The continuation of countering violent extremist radicalisation is required with the integrated help of authorities such as the CTIRU and social media platform regulators with the heuristic aim to decrease violent extremist behaviours and terrorist acts. Although it will not eliminate terrorism entirely, it will be a huge contribution. The idea of radicalisation is very complex and its diver-sity ranges from case to case in terms of how individuals are radicalised. Critiques assume radicalisation is wholly assumption and intuition based as it is not scientific empirically based research. This area of violent extremism requires more research for it to be deemed empirical evidence which could consequently reduce the risks of terrorist acts occurring in the future.

References

1. Ali M (2015) ISIS and propaganda: how ISIS exploits women. Reuters Inst Study Journal 10–11
2. Al-Rawi A (2018) Video games, terrorism, and ISIS's Jihad 3.0. Terror Polit Violence 30(4):740–760
3. Aly A, Macdonald S, Jarvis L, Chen TM (2017) Introduction to the special issue: terrorist online propaganda and radicalization
4. Amedie J (2015) The impact of social media on society
5. Anglano C, Canonico M, Guazzone M (2017) Forensic analysis of telegram messenger on android smartphones. Digit Investig 23:31–49
6. Anumala H, Busetty SM (2015) Distributed device health platform using Internet of Things devices. In: 2015 IEEE International conference on data science and data intensive systems, pp 525–531
7. Archetti C (2013) Terrorism, communication, and the media. In: Understanding terrorism in the age of global media, pp 32–59
8. Baruah TD (2012) Effectiveness of social media as a tool of communication and its potential for technology enabled connections: a micro-level study. Int J Sci Res Publ 2(5):1–10
9. Baugut P, Neumann K (2020) Online news media and propaganda influence on radicalized individuals: findings from interviews with Islamist prisoners and former Islamists. New Media Soc 22(8):1437–1461
10. Baumeister RF, Leary MR (1995) The need to belong: desire for interpersonal attachments as a fundamental human motivation. Psychol Bull 117(3):497
11. Benigni MC, Joseph K, Carley KM (2017) Online extremism and the communities that sustain it: detecting the ISIS supporting community on Twitter. PLoS ONE 12(12):e0181405
12. Bertram L (2015) How could a terrorist be de-radicalised? Jo Deradic 5:120–149
13. Bertram L (2016) Terrorism, the Internet and the social media advantage: exploring how terrorist organizations exploit aspects of the Internet, social media and how these same platforms could be used to counter-violent extremism. J Deradic 7:225–252
14. Bjelopera JP (2012) The domestic terrorist threat: background and issues for congress
15. Bloom M, Tiflati H, Horgan J (2019) Navigating ISIS's preferred platform: Telegram1. Terror Polit Violence 31(6):1242–1254
16. Blum A (2007) Machine learning theory. Carnegie Melon University, School of Computer Science, p 26
17. Borum R (2014) Psychological vulnerabilities and propensities for involvement in violent extremism. Behav Sci Law 32(3):286–305
18. Butler AS, Panzer AM, Goldfrank LR (2003) Understanding the psychological consequences of traumatic events, disasters, and terrorism. In: Preparing for the psychological consequences of terrorism: a public health strategy
19. Christensen G (2019) Artificial Intelligence of Things (AIoT). IoT agenda. https://internetofth ingsagenda.techtarget.com/definition/Artificial-Intelligence-of-Things-AIoT
20. Christmann K (2012) Preventing religious radicalisation and violent extremism: a systematic review of the research evidence
21. Cohen-Almagor R (2005) Media coverage of terror: troubling episodes and suggested guidelines. Can J Commun 30(3):383–409
22. Conway M, Dillon J (2019) Future trends: live-streaming terrorist attacks? VOX-Pol
23. Counter Terrorism Policing (2018) Specialist unit tackles online extremism. https://www.cou nterterrorism.police.uk/specialist-unit-tackles-online-extremism/
24. Crenshaw M (1981) The causes of terrorism. Comp Polit 13(4):379–399
25. Crosby F (1976) A model of egoistical relative deprivation. Psychol Rev 83(2):85
26. D'Souza SM (2015) Online radicalisation and the specter of extremist violence in India. Mantraya Brief (1)
27. Dekel R, Nuttman-Shwartz O (2009) Posttraumatic stress and growth: the contribution of cognitive appraisal and sense of belonging to the country. Health Soc Work 34(2):87–96

28. Diener E (1979) Deindividuation, self-awareness, and disinhibition. J Pers Soc Psychol 37(7):1160

29. D'Incau F, Soesanto S (2017) Countering online radicalisation [Blog]. https://ecfr.eu/article/commentary_countering_digital_radicalisation_7216/

30. Donovan J, Lewis B, Friedberg B (2018) Parallel ports. Sociotechnical change from the Alt-Right to Alt-Tech. In: Post-digital cultures of the far right. transcript-Verlag, pp 49–66

31. Doosje B, Moghaddam FM, Kruglanski AW, De Wolf A, Mann L, Feddes AR (2016) Terrorism, radicalization and de-radicalization. Curr Opin Psychol 11:79–84

32. Doosje B, van den Bos K, Loseman A, Feddes AR, Mann L (2012) "My in-group is superior!": susceptibility for radical right-wing attitudes and behaviors in Dutch youth. Negot Confl Manag Res 5(3):253–268

33. Douglas K (2010) Deindividuation. Encyclopaedia Britannica. https://www.britannica.com/topic/deindividuation#ref310686

34. Facebook Community Standards (n.d.) Objectional content. Hate speech. Community Standards (facebook.com)

35. Ganor B (2002) Defining terrorism: is one man's terrorist another man's freedom fighter? Police Pract Res 3(4):287–304

36. Gaydhani A, Doma V, Kendre S, Bhagwat L (2018) Detecting hate speech and offensive language on twitter using machine learning: an n-gram and tf-idf based approach. arXiv:1809.08651

37. Gill P (2007) A multi-dimensional approach to suicide bombing. Int J Confl Violence (IJCV) 1(2):142–159

38. Gill P, Horgan J, Deckert P (2014) Bombing alone: tracing the motivations and antecedent behaviors of lone-actor terrorists. J Forensic Sci 59(2):425–435

39. Gillis A (2020) Internet of things. IoT agenda. What is IoT (Internet of Things) and how does it work? (techtarget.com)

40. Goldberg ME, Gorn GJ (1987) Happy and sad TV programs: how they affect reactions to commercials. J Consum Res 14(3):387–403

41. Goodwin MJ (2011) New British fascism: rise of the British National Party. Routledge

42. Gray DH, Head A (2009) The importance of the Internet to the post-modern terrorist and its role as a form of safe haven. Eur J Sci Res 25(3):396–404

43. Grusin R (2010) Premediation: affect and mediality after 9/11. Springer

44. Guelke A (2009) The new age of terrorism and the international political system. IB Tauris

45. Hardy K (2018) Comparing theories of radicalisation with countering violent extremism policy. J Deradic 15:76–110

46. Hariman R, Lucaites JL (2007) No caption needed: iconic photographs, public culture, and liberal democracy. University of Chicago Press

47. Hasim NNM, Mohamed H, Ibrahim J (2016) The effect and challenges of online radicalization on modern day society. Int J Inf Commun Technol 6(12)

48. Hogg MA (2020) Uncertain self in a changing world: a foundation for radicalisation, populism, and autocratic leadership. Eur Rev Soc Psychol 1–34

49. Hogg MA, Kruglanski A, Van den Bos K (2013) Uncertainty and the roots of extremism. J Soc Issues 69(3):407–418

50. Home Office (2011) The prevent strategy. Crown Publishing Group, London, pp 107–108

51. Hudson RA (1999) The sociology and psychology of terrorism: who becomes a terrorist and why? Federal Research Div., Library of Congress, Washington DC

52. Hussin S (2018) Singapore's approach to countering violent extremism. Combatting violent extremism and terrorism in Asia and Europe. Konrad Adenauer Stiftung, p 171

53. Jones S (2009) Radicalisation in Denmark. Renewal 17(1):22–28

54. Jones S (2018) Radicalisation in the Philippines: the Cotabato Cell of the "East Asia Wilayah." Terror Polit Violence 30(6):933–943

55. Jowett GS, O'donnell V (2018) Propaganda & persuasion. Sage Publications

56. Koehler D (2014) The radical online: individual radicalization processes and the role of the Internet. J Deradic 1:116–134

57. Kovács A (2015) The 'new jihadists' and the visual turn from al-Qa'ida to ISIL/ISIS/Da'ish. Bitzpol Aff 2(3):47–69
58. Krona M (2020) Collaborative media practices and interconnected digital strategies of Islamic State (IS) and Pro-IS supporter networks on telegram. Int J Commun 14:1888–1910
59. Kundnani A (2012) Radicalisation: the journey of a concept. Race Class 54(2):3–25
60. Macdonald S, Correia SG, Watkin AL (2019) Regulating terrorist content on social media: automation and the rule of law. Int J Law Context 15(2):183–197
61. McCauley C, Moskalenko S (2008) Mechanisms of political radicalization: pathways toward terrorism. Terror Polit Violence 20(3):415–433
62. Milmo C (2014) Iraq crisis exclusive: Isis jihadists using World Cup and Premier League hashtags to promote extremist propaganda on Twitter. The Independent
63. Moghaddam FM (2008) How globalization spurs terrorism: the lopsided benefits of "one world" and why that fuels violence. Praeger Security International
64. Moghaddam FM, Heckenlaible V, Blackman M, Fasano S, Dufour DJ (2016) Globalization and terrorism. Soc Psychol Good Evil 415
65. Neo LS, Dillon L, Khader M (2017) Identifying individuals at risk of being radicalised via the Internet. Secur J 30(4):1112–1133
66. Patrikarakos D (2018) Web 2.0: the new battleground. Armed Confl Surv 4(1):51–64
67. Radicalisation Awareness Network (2017) Working with families and safeguarding children from radicalisation. https://bit.ly/2UEP29a
68. Ramakrishna K (2011) Self-radicalisation and the Awlaki connection. In: Strategic currents. ISEAS Publishing, pp 140–142
69. Sageman M (2004) Understanding terror networks. University of Pennsylvania Press
70. Sageman M (2008) A strategy for fighting international Islamist terrorists. Ann Am Acad Pol Soc Sci 618(1):223–231
71. Salahuddin M, Alam K (2015) Internet usage, electricity consumption and economic growth in Australia: a time series evidence. Telemat Inform 32(4):862–878
72. Schmid AP (2013) Radicalisation, de-radicalisation, counter-radicalisation: a conceptual discussion and literature review. ICCT Res Paper 97(1):22
73. Schroeter M (2020) Global network on extremism & technology. Artificial intelligence and countering violent extremism: a primer. GNET-Report-Artificial-Intelligence-and-Countering-Violent-Extremism-A-Primer_V2.pdf
74. Silber MD, Bhatt A, Analysts SI (2007) Radicalization in the West: the homegrown threat. Police Department, New York, pp 1–90
75. Smith BL (2020) Propaganda. Encyclopaedia Britannica. Propaganda | Definition, history, techniques, examples, & facts | Britannica
76. Spears R, Lea M, Lee S (1990) De-individuation and group polarization in computer-mediated communication. Br J Soc Psychol 29(2):121–134
77. Speckhard A, Ellenberg M (2020) Is Internet recruitment enough to seduce a vulnerable individual into terrorism. Homeland Security Today
78. Statista (2020) Global digital population as of October 2020 [Graph]. https://www.statista.com/statistics/617136/digital-population-worldwide/
79. Stewart BB, Thompson JW (2002) U.S. Patent No. 6,414,635. U.S. Patent and Trademark Office, Washington, DC. 1498394371873592871-06414635 (storage.googleapis.com)
80. Striegher JL (2015) Violent-extremism: an examination of a definitional dilemma
81. Suler J (2004) The online disinhibition effect. Cyberpsychol Behav 7(3):321–326
82. Tajfel H, Turner J (1986) The social identity theory of intergroup behavior. Psychology of intergroup relations
83. Terrorism Act 2000 (TACT) https://www.legislation.gov.uk/ukpga/2000/11/section/1#:~:text= (1)In%20this%20Act%20%E2%80%9C,section%20of%20the%20public%2C%20and
84. Thomas TL (2003) Al Qaeda and the Internet: the danger of 'cyber planning'. Foreign Military Studies Office (ARMY) Fort Leavenworth Ks
85. Torok R (2013) Developing an explanatory model for the process of online radicalisation and terrorism. Secur Inform 2(1):6

86. Trip S, Bora CH, Marian M, Halmajan A, Drugas MI (2019) Psychological mechanisms involved in radicalization and extremism. A rational emotive behavioral conceptualization. Fronti Psychol 10:437
87. Tzezana R (2016) Scenarios for crime and terrorist attacks using the Internet of Things. Eur J Futures Res 4(1):18
88. United Nations Office on Drugs and Crime (2012) Use of the Internet for terrorist purposes
89. Van den Bos K (2018) Why people radicalize: how unfairness judgments are used to fuel radical beliefs, extremist behaviors, and terrorism. Oxford University Press
90. Villasenor J (2020) How to deal with AI-enabled disinformation. Brookers
91. Vincent J (2019) Facebook bans UK's biggest far-right organizations, including EDL, BNP, and Britain First [Blog]. https://www.theverge.com/2019/4/18/18484623/facebook-bans-uk-far-right-groups-leaders-edl-bnp-britain-first
92. Weimann G (2004) www.terror.net: how modern terrorism uses the Internet, vol 31. United States Institute of Peace
93. Weimann G (2006) Virtual disputes: the use of the Internet for terrorist debates. Stud Confl Terror 29(7):623–639
94. Weimann G (2014) New terrorism and new media, vol 2. Commons Lab of the Woodrow Wilson International Center for Scholars, Washington, DC
95. Weimann G (2016) Going dark: terrorism on the dark web. Stud Confl Terror 39(3):195–206
96. Wibisono S, Louis WR, Jetten J (2019) A multi-dimensional analysis of religious extremism. Front Psychol 10:2560
97. Wiktorowicz Q (ed) (2004) Islamic activism: a social movement theory approach. Indiana University Press

Exploring the Extent to Which Extremism and Terrorism Have Changed Since the Advent of the Internet

Kevin Locaj

Abstract This review will examine the academic literature over which role the internet has in the evolution of extremism and terrorism since its advent. It will compare two different approaches. The first claims that the internet is a major factor that facilitates ideas and narratives, which can lead to the rise of extremism and terrorism. The second, which in its turn contradicts this approach, argues that prior the advent of the internet extremists and terrorists where more successful into achieving their goals. For that reason, the review will be split into 3 sections. The first section will be examining some needed key definitions of what constitutes terrorism and extremism. Afterwards the essay will shift its approach towards the main debate of whether the internet has a causal link with extremism and terrorism or not. Therefore Sect. 2 brings forth the ways, in which the internet has helped terrorism to advance its goals. Moving on to the third section, this piece of work will discuss the approach in which the internet does not assist extremist narratives but, in the contrary helps the advancement of better research around it and its prevention. Lastly, the review will sum up over the literature that has been discussed and conclude that there is always space for future research.

Keywords The Internet · Cyberspace · Extremism · Terrorism · Propaganda · Radicalization

1 Introduction

Since the advent of the internet in people's lives, many things have changed in terms of socialization, communication, trade, finances etc. "In today's connected world, communications have never been faster, more convenient or more prolific". Its ubiquity in the average person's life has made it one of the most studied things

K. Locaj (✉)
Hillary Rodham Clinton School of Law, Swansea University, Singleton Park, Swansea SA2 8PP, UK
e-mail: kevinloc94@yahoo.com
URL: http://www.swansea.ac.uk

© The Author(s), under exclusive license to Springer Nature Switzerland AG 2023
R. Montasari et al. (eds.), *Digital Transformation in Policing: The Promise, Perils and Solutions*, Advanced Sciences and Technologies for Security Applications,
https://doi.org/10.1007/978-3-031-09691-4_8

137

nowadays [1]. One thing that also is claimed to have change is the way which the internet affects extremism and terrorism. More specifically, there is a concern among academics and counterterrorism policymakers, about how easy a violent extremist content can be accessible online and whether that content can create acts of terrorism [8]. However, claims that there is not enough proof available to associate violent extremist and terrorist engagement, with the consumption of extremist content that can be found online [8]. After all, the internet does not apply many restrictions towards the spread of extremist content from the producer towards the consumer along with social interaction around it. "It is precisely the functionalities of the social Web that causes many […] to believe that the Internet is playing a significant role in contemporary radicalization processes" [8]. Thus, a large number of violent extremists who may have been influenced by online preachers and terrorist propagandists, has shifted the attention of scholars, counterterrorism policymakers towards the influence which the internet might have to these people and at what level it has "radicalized" them into committing terrorist actions [1].

Although most academics are not able to adequately answer whether the internet has a negative or a positive linkage between contemporary extremism and terrorism, many studies attempt to examine the factors which determine whether an individual will sympathize or not, with terrorist propaganda just by consuming its content online. In spite of the contemporary concern about the internet having a role in violent extremism and terrorism, the scepticism around the matter is not new. Rapoport [22], in his wave theory of terrorism, determines that the way technology is being used, and more specifically the communication technologies that are being used, can influence the spread, type and timing of extremist and terrorist content [22]. Moreover, as this digital revolution has dramatically changed people's lives worldwide, has become indistinguishable to tell which is the "online" and which is the "offline" [33]. As a matter of fact the internet is not new for many terrorists. According to Weimann [31], terrorists are considered to be some of the first adopters of the internet. Internet has been for long time been a major link between extremism, whether in its violent form or not [25, 31]. As it is clear, the internet has become essential for terrorist across the globe. It can be used to serve their purposes, such as planning, propaganda as well as financing and recruitment. It has become their primary facilitator which can provide access to a large number of audiences so they can express their political and religious ideologies [33]. There is no doubt that the internet will remain a key factor for the cause of extremism and terrorism for years to come. The question is, at what level the internet will influence such ideas and also in what level can help into mitigating them? As history shows terrorists adapt to the technology that advances around them, but so do the counterterrorism policy makers who can benefit from the internet in a more efficient way. Indeed, although terrorists might be constant users of the internet, it cannot be claimed that they use it in the most efficient way. This leads to the assumption that governments have the upper hand in this instance.

1.1 Defining Terrorism

To this date there is no universal unanimity over the definition of what constitutes terrorism. Terrorism is a phenomenon that happens on a local level as well as on transnational. Thus, terrorism becomes complicated and hard to define. Hence, "there is no universally agreed definition" and thus, a variety of definitions exists and much confusion as what constitutes it [27]. As Ljujic [18] pinpoints, there are many formidable, punishable and illegal acts, but a vast majority of people perceive terrorist acts as one of the most formidable threats to national security [18]. It is an act of crime that indiscriminately can target civilians. Moreover the UK defines terrorism as "the use or threat of action, both in and outside of the UK, designed to influence any international government organisation or to intimidate the public. It must also be for the purpose of advancing a political, religious, racial or ideological cause." The actions can also include "serious violence against a person or damage to property, endangering a person's life (other than that of the person committing the action) creating a serious risk to the health or safety of the public or a section of the public, action designed to seriously interfere with or seriously to disrupt an electronic system" [28]. Furthermore, Crenshaw's approach of terrorism perceives it as "a form of political behaviour resulting from the deliberate choice of a basically rational actor, the terrorist organization" [10]. Nevertheless, she stretches out, that there are also many factors to be considered—social, economical, political conditions—which can raise the probability of terrorism to occur in some places rather than others. Therefore, terrorism is a complicated concept and it is defined in this section as a punishable crime, designed to influence and spread fear, in which the actor can be an individual or an organization, where indiscriminately uses violence or threatens to do so, against a person or property, motivated in its majority from political and religious purposes.

1.2 Defining Extremism

As far as there is discussion around the matter of what constitutes extremism, there is also no consensus among the scholars, as such complicated term truly demands constant and comprehensive study. The word on its own is a relative term. According to Schmid [24]. It always demands a benchmark to define what constitutes the "normal", "ordinary" the "mainstream" when it has to be compared with the "extreme" [24]. It is true that a society's conceptions for what constitutes the 'norm' change over the time. Hence, the conception of what constitutes the 'extreme' also changes. For that reason, most academics avoid to give a unique definition for extremism and therefore, as Neumann [21] states, they decide to adopt a more flexible approach "that can be used in the context of both actions ("behavioural radicalization") and beliefs ("cognitive radicalization") [21]. In other words, Neumann states that someone can hold extremist ideas, without these ideas necessarily leading him into extremist actions [21]. According to Wibtrope [32], an individual and a group can be categorized into 3 categories based on how they decide to fulfil their extreme objectives. There are those, "that their objectives are extreme and their means to achieve them are alike, those that may have extreme objectives but decide not to use extreme means; and finally there are who have negotiable objectives but they use extreme means to fulfil them" [32]. This assertion in spite of giving a more flexible approach of who should be considered as an extremist falls short in terms of contemporary standards around the term. In other words, according to Wibtrope's categorization, someone who is under the second category is considered to be a non violent extremist if compared to the first category. Hence, extremism in this case, is defined by the level of someone's commitment to take extreme action. While this categorization between doctrinal and functional extremism might be useful, it should not be the definitive one. Finally, Winter [33] tries to define specifically online terrorism as "Internet activism that is related to, engaged in, or perpetrated by groups or individuals that hold views considered to be doctrinally extremist". Therefore, this research will define extremism, as activism that can be engaged in, or perpetrated by individuals and groups, it can be doctrinal or functional, and the actions should be viewed as such, based on what a particular society has set as benchmark of what constitutes 'extreme'.

2   The Internet as an Extremist and Terrorist Facilitator

Over the last decade it has been asserted that the internet is responsible for fishing and potentially recruiting active extremists among Western populations [6]. This assertion although explains a specific circumstance in which people get influenced from a certain usage of a technology. Yet it is an oversimplified view, that a technology, just by simply existing must produce certain effects. It completely overlooks the fact that people usually use this technology for their own benefit and prosperity. It is always a matter of choice, to actively adopt extreme narratives rather than 'merely absorbing' an idea that simply exists [2]. Contrary to that, Timothy [29] claims that the internet is a major contributor for terrorists to plan their attacks [29]. Indeed terrorists according to Winter can use the internet for purposes such as recruiting etc. [33]. Moreover, Helfstein [13] argues that the internet can create 'leaderless' organizations because it allows terrorists to act more independently [13]. In addition as the internet has become an essential feature in modern life, it sparks fears among scholars and counterterrorism, that this new technology can be a helpful tool to breed terrorism [5]. Indeed the advent of the internet has revolutionized the way people communicate and interact, therefore it is not unreasonable for one to claim that it has also revolutionized the way that terrorists fight their wars. This section will review in what ways the internet can facilitate some individuals in terms revolutionizing their tactics. It also reviews the most common use of it by terrorists and extremists, on either individual or organizational level.

2.1 Recruitment

One of the most important parts of a terrorist organization is to convince someone to join their cause. According to Conway [9] who examined the role which the charisma might have in to potentially recruiting members in terrorist organization, she shifted her attention into the online recruitment methods which are being used by Al Qaeda in Arab Peninsula (AQAP). She argues that the charisma alone is not enough to influence an individual and create an extremist environment around him, let alone to bait him into getting involved in a terrorist group. She pinpoints that a face to face interaction is necessary in order for this to succeed [9]. What she also notes, is that the online influence which AQAP might try to achieve around the Arab Peninsula, is not as successful as in other parts of the world [9]. That being said, recruitment of an individual into a terrorist organization might facilitated by the internet, but sometimes it is at what extension people can afford to use this technology that counts. Hence, AQAP might not be as successful into influencing the audience they aim and therefore, they cannot recruit individuals from the Arab Peninsula as much as from other parts of the world.

2.2 Propaganda

It is understandable among many scholars why many terrorist organizations will choose propaganda as a strategy to draw sympathizers. In their study, Mozes and Weimann [19] investigating the Palestinian Islamist organization of 'Hamas' they found that 'Hamas' used the internet to successfully facilitate and spread their ideas [19]. More recently, the IS propaganda has drawn the attention of the media and therefore academics' along with policymakers' focus for further investigation. This is due to the fact that the media publish by far, more of their content and thus it has raised public policymakers' caution. Moreover, Holbrook [14] in his study on both far-right and Islamist propaganda compares the two types of activism in order to analyze the similarities and the differences so it can be understood how far-right is born out of reaction against Islamist activism [14]. Indeed, Islamist propaganda has sparked concerns around the rise of far-right movements that claim to be anti-terrorist. Nevertheless, the internet offers cheap and accessible extremist propagandist content to numerous potential sympathizers, no matter their position on the map might be. Furthermore, as Sheikh's [26] study claims, the online terrorist propaganda "not only did help recruit new adherents, it also kept the organization's ranks coherent [26]. Finally, as Atwan [3] along with Nacos [20] claim, terrorist are using the internet through the publication of 'glossy' magazines, Hollywood-style films, and depicting warfare action sequences, in that way these groups aim to communicate with potential sympathizers. The jihadist group ISIS for example, has been found to be extremely successful online, it has masterfully used the internet in order to spread its propagandistic materials in pursuit of the potential earning of financial and military support [3, 20]. With the advent of internet, terrorist organizations can prepare their messages for distribution in near no time. Indeed, there is a plethora of small sized videos of such content one can find online and they are designed perfectly for a large audience. Terrorist organizations now have direct access to the populace they aim to influence or terrorize.

2.3 *Funding and Logistics*

A number of studies have been conducted into what level online technology is being used from terrorist groups to facilitate terrorism. Hughes along with Meleagrou-Hitchens [15] holds that there is a complicated logistical system which includes fighters who motivate, encourage and instruct potential candidates to join their fights, from the 'West' [15]. This material is accessible online for people to sympathize with those people's struggle and get motivated into joining their ranks. The material is usually professionally set in a certain way in order to inspire candidates outside of the countries in which these groups are active. Thus, the content that is available online has two goals. The first goal is to instruct and offer a logistical advice, and the second is to legitimize and justify their cause, in order to inspire future extremist sympathizers so they can reinforce their ranks [23]. Another side that the internet can facilitate in favour of the spread of extremism and terrorism is by helping groups getting financial resources. Even though analysts tend to turn their attention towards the propagandist material and the way of online recruitment is conducted, Jacobson [16] pinpoints that are also illicit transactions that are conducted online by terrorist groups, with more ease than ever, which usually finance their attacks [16]. As it is clear, such causes usually are armed attacks. A respected amount of capital is usually needed and terrorists use the internet for that matter in two ways. The first, as it is aforementioned to conduct illicit transactions and purchase weaponry in black markets and the second is to create fundraisers, so potential sympathizers and supporters to contribute in their cause [33]. As Camstoll Group [7] specifically cites, "al-Qaida and ISIS fundraisers have taken credit for millions of dollars raised using social media-based campaigns" [7]. Moreover, terrorist financing is moving slowly towards the cryptocurrency meaning that one day in the future 'cryptos' may facilitate extremist activists [12]. Although this assertion might be based on mostly anecdotal evidence, yet it highlights flaws on the financial system that in the future can get exploited.

2.4 Anonymity, Accessible Information, Cheap Communication

There are three more ways that the internet can facilitate extremism and terrorism. It can provide relatively easy anonymity to its users compared to other tools that might have been used in the past. It also gives, as it is aforementioned, relatively easy access to an abundance of information that might be useful to terrorists. Finally, it provides its users cheap with a communication tool, which can connect two or more individuals in an instant no matter their geographical position. It is well known that the internet can easily provide anonymity; fake profiles and stories with absolutely no verification can be found online by almost anyone nowadays [5]. Thus, a terrorist through the anonymity which the internet can provide to him can inflict serious damage without risking himself in the process. Despite anonymity giving a relative impunity to a terrorist, this is not always their desire. In other words, if an extremist narrative needs to get delivered by a group, anonymity provides near to nothing in terms of communication with the audience it aims to influence or terrorize. One the other hand, what a group might lose in terms of communication, it benefits from not getting connected with certain violent actions, meaning they can keep their propaganda wing untouched [5]. Additionally, the internet has made available some information which grants the ability to individuals terrorize more efficiently. This ability is to violently act in a more effective and professional way against human lives and properties and this information used to be under the states' control in the past [15]. It seems nowadays that this information, which in the past usually was practiced by people in the military, now, one can easily have online access to it. As Benson (2014) quotes, "The virtual jihadist network has replaced al Qaeda training camps" (Benson, 2014). Moreover, terrorists in the past had long distances to separate them, while now they are only a 'click' away from one another. The internet through the social platforms has drastically decreased the cost of communications. Thus, an extremist group has now the luxury, to communicate in real time with its members and sympathizers relatively cheap.

3 The Internet Not Facilitating Extremism and Terrorism

Despite the aforementioned section seems to conclude that the internet has become a very effective tool in the hands of violent extremists, many scholars have argued that this is not always the case. As Conway [8] argues, there is an abundance of violent extremism content online, but not all consumers are affected in the same way [8]. In addition, Frissen [11] in his study is unable to conclude that the internet is the main factor to draw an individual towards radical ideologies, but rather someone's willingness to search for extremist material online so they come to adopt it in the end [11]. Furthermore, negative arguments of the role of the Internet and social media are also based on a lack of historical perspective. As Archetti [2] holds that, what might be perceived as 'communication revolution' is barely the latest evolution of any communication technology that existed before, from the invention of the telegraph until now, technology has always been used to benefit the lives of people, as well as for malicious reasons [2] [33]. Although the concerns about the internet being a terrorist facilitator are not unjustified, they tend to shift the attention from other factors that might be the cause extremist ideologies. In fact the internet facilitates many other political actors (e.g. NGOs, governments, academic, etc.), if not more, it at least facilitates them as much as it does with terrorists. Therefore, the terrorist organizations are not the only ones that benefit from the advent of the internet and thus official states can use the internet in order to advance their political agendas and also counter violent extremism. This section will be discussing the argument which claims that the internet is more beneficial to governments and thus it becomes harder for terrorists to use it against them [5, 31, 2].

3.1   Government Superiority on the Internet

Taking into account the advantages the internet has to offer, one can easily argue that the governments have a greater benefit from the use of it, rather than the terrorists. Indeed, states not only are superior in terms of knowledge and equipment, they also can use the advantages the terrorists think they have, against them. According to Benson [5] government can have access the same amount of information that exists online without having the fear of surveillance [5]. In other words, a counterterrorist can conduct research about the ways extremists navigate through the internet and learn how to counter them without having the fear of prosecution for his action. Also, governments are enormous organizations in comparison with a small terrorist group or even if it is compared with the biggest terrorist organizations, governments overpower them in almost every aspect, with thousands of employees in their ranks. In addition, governments do not need to rely on online propaganda campaigns to recruit people and also as Benson [5] cites, they "do not rely on YouTube to learn how to conduct a raid, hack a computer, or build a bomb" [5, 2].

3.2   The Internet's Anonymity Not Being Standard

There is the assumption that just because somebody can create a fake account or fake profile online, this gives him the adequate anonymity to navigate online without him being tracked. Remaining anonymous online cannot be achieved for long time due to the fact of that is impossible not to leave behind information that acts like 'track on the snow' [5, 30]. The internet is constructed in such way that, the information shared online, is always recorder from its source to its receiver. Thus, when an information is uploaded on the web several organizations who own search engines and also social media platforms, will download and archive that information, meaning that even someone attempts to delete it, it will in fact still exist [5, 11]. According to Benson [5] "Companies like Facebook [...] also have a history of keeping deleted data for the convenience of returning customers and even gathering data without the user's knowledge" [5]. A fine example of the fact that your actions can be tracked online is the case where the infamous organization "Anonymous" tried to hack into individuals and organizations. In spite of their perceived anonymity, many the organization's perpetrators got prosecuted. Finally, another disadvantage online anonymity has for the violent extremists, is that part of their success is defined by the trust there needs to exist among the members of the organization. As Frissen [11] and Benson [5] argue, a terrorist must identify a potential online sympathizer beyond his hidden identity in order to keep the integrity of the group against outside infiltration [5, 11]. In that case there is a chance for counter terrorists to identify such people, and in fact in a more efficient way, due to their superiority of resources.

3.3   Online Extremist Information

The online world offers a plethora of information for its users to consume in almost no cost. Among this information, there is an abundance of extremist material and propaganda which individuals either decide to absorb or they search it for research purposes. Moreover, Benson [5] argues that terrorist groups support individuals who are willing to create content online, in order to advance their goals [5]. Additionally, the fact that material exist online does not mean that it is easy to be acquired by an individual. In other words, as Frissen [11] along with Benson [5] and Archetti [2] claim, in order to assess something's integrity, the source must be known to the receiver. Thus, the condition of anonymity is lost for someone who wishes to upload extremist material without him drawing the attention of the authorities. Furthermore, as it is aforementioned, one can find online information about how to acquire some particular skills in order to get involved into violent extremism. Yet, the fact that this kind of material is just accessible does not necessarily mean that is there to be acquired by anyone. As Benson [5] simply puts it, an instructor is usually needed for the adequate training of a particular set of skills that someone wishes to acquire. "Recipes and instruction on gourmet cooking are widely available [...] yet despite the benefit of so much information, many people remain unable to prepare food correctly, let alone master gourmet cooking" [5]. Of course, this does not mean that the danger of a self-taught online extremist is eliminated in any way, it is just really difficult to be achieved just because it is accessible. In addition to those, the process of the absorption of such content is more complicated. Humans, has been argued to have their own a moral compass which guides them towards a universal standard way of behaviour [4]. People cognitively will decide whether to sympathize or not with extremist ideology. In other words some material found online, might have the opposite results of those it was aiming for. For example beheading videos has been found to have little influence into drawing supporters towards extremist ideology [17]. Hence, sometimes the internet is not always beneficial in the hands of terrorists, especially if there is an absence of adequate knowledge behind to support it.

4 Conclusion

In conclusion, there is no possible way to predict whether terrorism will rise through the advent of the internet or not. Undoubtedly, the internet has played a significant role into facilitating and advancing the goals of extremism and terrorism, which might have been preached individually or at an organizational level. The advent of the internet has created an unprecedented advancement in the communications sector, which puts terrorist narratives in position, to spread at rapid and easy level and without considering boundaries. Thus, obstacles of the past are over passed with lower than ever cost, and hence, terrorist sympathizers are now able to communicate with each other almost instantly despite their geographical position. It also offers them the availability of acting, most of the times secretly, without facing almost any repercussions. These luxuries which the internet offers nowadays create understandable concerns around the fears for a rise in terrorism and extremism. Contrary to these benefits which the internet has to offer to extreme activists, there are arguments which state that, the internet is helping to prevent and reduce terrorism rather than facilitating it. Indeed, over the last decades the internet has allowed advancement in sectors such as communications, economy and security. Counter terrorism policies have now a powerful tool in their hands which allows them to coordinate their actions against the rise of terrorism. Moreover, researchers of the subject are being able to look into the matter deeper, in order to understand its roots and complexity. In addition, the benefits the internet has to offer for terrorists, might not be as useful to them as many people fear. The abundance of information that lies in there, it might not be hard to obtain, but it can be difficult to be used efficiently. On top of that, the sense of anonymity which the internet offers is mostly a deception and in some cases it does not even serve their purposes. There is no doubt that the internet has played a crucial role into facilitating extremist and terrorist ideologies, but on the other hand it has benefited the research on the subject, as well as the better understanding of its roots while at the same time it helps preventing it. Finally, whether extremism and terrorism are facilitated, or not, from the internet are conclusions that do not constitute a panacea and they cannot be generalized. Different trends usually exist, but there are no universal structures or systems, thus, any academic approach must take a more specific look in order to understand them, and therefore fragments of different spheres of activity must be created in order to address the problem in a more efficient way.

References

1. Aly A, Macdonald S, Jarvis L, Chen T (2016) Introduction. In: Aly A, MacDonald S, Jarvis L, Chen T (eds) Terrorism online: politics, law and technology. Routledge, Abingdon, Oxon, pp 1–7
2. Archetti C (2015) Terrorism, Communication and New Media: explaining radicalization in the digital age, *Perspectives On Terrorism,* 9(1), 49–56
3. Atwan AB (2015) Islamic state: the digital caliphate. University of California Press
4. Bandura A (2002) Selective moral disengagement in the exercise of moral agency. J Moral Educ 31(2):101–119. https://doi.org/10.1080/0305724022014322
5. Benson DC (2014) Why the Internet is not increasing terrorism. Secur Stud 23(2):293–328. https://doi.org/10.1080/09636412.2014.905353
6. Brooking E (2014) The ISIS propaganda machine is horrifying and effective. How does it work? Council on Foreign Relations Blog. http://blogs.cfr.org/davidson/2014/08/21/the-isis-propaganda-machine-is-horrifying-and-effective-how-does-it-work/. Accessed 20 Dec 2021
7. Camstoll Group (2016) Use of social media by terrorist fundraisers and financiers. Los Angeles and Washington, D.C. https://www.camstoll.com/wpcontent/uploads/2016/04/Social-Media-Report4.22.16.pdf. Accessed 29 Nov 2021
8. Conway M (2017) Determining the role of the Internet in violent extremism and terrorism: six suggestions for progressing research. Stud Confl Terror 40(1):77–98. https://doi.org/10.1080/1057610X.2016.1157408
9. Conway M (2012) From Al-Zarqawi to Al-Awlaki: the emergence of the internet as a new form of violent radical milieu
10. Crenshaw M (1981) The causes of terrorism. Comp Polit 13(4):379–399. https://doi.org/10.2307/421717
11. Frissen T (2021) Internet, the great radicalizer? Exploring relationships between seeking for online extremist materials and cognitive radicalization in young adults. Comput Human Behav 114:1–10. https://doi.org/10.1016/j.chb.2020.106549
12. Goldman ZK, Maruyama E, Rosenberg E, Saravalle E, Solomon-Strauss J (2017) Terrorist use of virtual currencies: containing the potential threat. Washington, DC. http://www.lawandsecurity.org/wp-content/uploads/2017/05/CLSCNASReportTerroristFinancing-Final.pdf. Accessed 29 Dec 2021
13. Helfstein S (2009) Governance of terror: new institutionalism and the evolution of terrorist organizations. Public Adm Rev 69(4):727–739
14. Holbrook D (2013) Far right and islamist extremist discourses: shifting patterns of enmity. In: Max Taylor PMC, Donald H (eds) extreme right wing political violence and terrorism, London: Bloomsbury Academic, 215–37
15. Hughes S, Meleagrou-Hitchens A (2017) The threat to the United States from the Islamic State's virtual entrepreneurs. Combat Terror Center Sentin 10(31):1–8
16. Jacobson M (2009) Terrorist financing on the internet. Combating terrorism center sentinel 2(6):17–20. https://www.washingtoninstitute.org/uploads/Documents/opeds/4a438817e3a3c.pdf.
17. Klausen J, Libretti R, Hung BWK, Jayasumana AP (2018) Radicalization trajectories: an evidence-based computational approach to dynamic risk assessment of 'homegrown' jihadists. Stud Confl Terror 1–28. https://doi.org/10.1080/1057610X.2018.1492819
18. Ljujic V, van Prooijen JW, Weerman F (2017) Beyond the crime-terror nexus: socio-economic status, violent crimes and terrorism. J Criminol Res Policy Pract 3(3):158–172. https://doi.org/10.1108/JCRPP-02-2017-0010
19. Mozes T, Weimann G (2010) The E-marketing strategy of Hamas. Stud Confl Terror 33(3):211–225
20. Nacos BL (2016) Mass-mediated terrorism. In: Mainstream and digital media in terrorism and counterterrorism. Rowman & Littlefield, Plymouth
21. Neumann P (2013) The trouble with radicalization. Int Aff 89(4):873–893

22. Rapoport D (2002) The four waves of rebel terror and September 11. Anthropoetics 8(1)
23. Reed A, Ingram HJ (2017) Exploring the role of instructional material in AQAP's inspire and ISIS's Rumiyah. Europol, The Hague. https://www.europol.europa.eu/publications-docume nts/exploring-role-ofinstructional-material-in-aqaps-inspire-and-isis-rumiyah. Accessed 03 Jan 2022
24. Schmid A (2014) Violent and non-violent extremism: two sides of the same coin? International Centre for Counter-Terrorism, The Hague. https://www.icct.nl/download/file/ICCT-Schmid-Violent-Non-Violent-Extremism-May2014.pdf. Accessed 15 Dec 2021
25. Seib P, Janbek DM (2010) High tech terror: Al Qaeda and beyond. In: Global terrorism and new media: the post-Al Qaeda generation. Routledge, Abingdon-on-Thames
26. Sheikh J (2016) "I just said it. The state": examining the motivations for Danish foreign fighters in Syria. Perspect Terror 10(6):3–11
27. Shinn D (2016) Poverty and terrorism in Africa: the debate continues. Georget J Int Aff 17(2):16–22. https://doi.org/10.1353/gia.2016.0020
28. Terrorism Act 2000 (2009) Part 1. https://www.legislation.gov.uk/ukpga/2000/11/section/1. Accessed 3 Jan 2022
29. Timothy LT (2003) Al Qaeda and the Internet: the danger of 'cyberplanning.' Parameters 23(1):112–123
30. Walker W, Conway M (2015) Online terrorism and online laws. Dyn Asymmetr Confl 8(2):156–175
31. Weimann G (2006) Virtual disputes: the use of the Internet for terrorist debates. Stud Confl Terror 29(7):623–639
32. Wibtrope R (2012) Rational extremism: the political economy of radicalism. Cambridge University Press, Cambridge
33. Winter C, Neumann P, Meleagrou-Hitchens A, Ranstorp M, Vidino L, Fürst J (2020) Online extremism: research trends in Internet activism, radicalization, and counter-strategies. Int J Confl Viol 14(2):1–20. https://doi.org/10.4119/ijcv-3809

Zero Trust Security Strategies and Guideline

Jim Seaman

Abstract Zero Trust is like the design of the layout of a high security military base, airport or bank, where the security is embedded into the design stage. Consequently, Zero-Trust strategies incorporate 'Secure by Design' principles into the early constructs of an organization's architecture. The premise of Zero-Trust (as the name indicates) is to assume that every interaction begins in an untrusted state. In contrast, traditional perimeter security often determines trustworthiness based on whether communication starts from inside a firewall. Zero-Trust is an alternative to traditional perimeter security, which due to the widespread moves to hybrid environments, remote working and insider risk has become increasingly regarded as being a flawed concept. Technology has evolved, as have work environments, but equally the cyber criminals' tactics and techniques have evolved as well. It's not a matter of if, but when, that opportunist malicious actor gets beyond your perimeter and organizations need to design their security architecture so that it predicts this. Zero trust starts by focusing on identifying those assets that need to be protected, by defining your protect surface (e.g. critical data, application, assets, and services), and then mapping the data and transaction flows across this (identifying any connected/impactful assets) and designing a zero-trust architecture through the effective siting of next-generation firewalls so that the defined protect surfaces are isolated and access is strictly restricted, based upon the principle of least privilege. Finally, understanding the environment your attack surfaces, needing protection, you are better able to understand what NORMAL is so that you can quickly find and respond to the detection of any potential **ABNORMAL** activities so that you can apply the 6 Ds of Defense (Deter, Detect, Disrupt, Delay, Deny and Defend).

Keywords Network architecture · Secure by Design · Secure by Default · Least privilege · Access control · Network security · Perimeter defense · Network segmentation · Perimeter security · Cloud security · Cybersecurity · Physical security

J. Seaman (✉)
IS Centurion Consultancy Ltd., Castleford, UK
e-mail: contact@iscenturion.com
URL: https://www.iscenturion.com

© The Author(s), under exclusive license to Springer Nature Switzerland AG 2023
R. Montasari et al. (eds.), *Digital Transformation in Policing: The Promise, Perils and Solutions*, Advanced Sciences and Technologies for Security Applications,
https://doi.org/10.1007/978-3-031-09691-4_9

1 Introduction

As we see an increasing number of reports for businesses that have become victims of cyber-attacks or unauthorized incursions, from malicious threat actors (which can have the potential to significantly increase, as the result of world events (e.g., Ukraine and Russia conflict [1]), there has been a desire by organizations to investigate alternative security models. Zero Trust (ZT) being one of the latest 'buzz terms' that is increasing in popularity, as companies seek ways to reduce their risks and their chances of becoming their enemy's next victim.

Traditional security models have relied on a robust perimeter (single point of failure) which appears to have proven to be ineffective in today's evolving digital business (e.g., Use of Artificial Intelligence) and especially for critical infrastructure (e.g., Protection of national infrastructure). However, the reality is that blaming the traditional security models could be a disservice to these security models, as the vulnerabilities might be caused by poor design, implementation and maintenance of these security models, rather than a flaw in these models.

Anyone that has served on a high security military base will fully appreciate the concept of Zero Trust Architecture. In such an establishment, despite you having a military identification card, this does not automatically give you the right to access anywhere on the military base. In fact, your military identification identity card might not even give you access to the base, or might limit your access to the perimeter and, perhaps, a handful of lower sensitivity buildings.

A Zero Trust model incorporates the principles of 'Secure by Design' and 'Secure by Default' into the development of any security architecture, be that network security or physical security. Rather than being focused on securing the perimeter, the Zero Trust model works on the presumption that the perimeter will be breached and that suitable measures need to be implemented to limit the movements from any unauthorized incursion. Consequently, the internal architecture is further compartmented and protected with robust authentication, authorization and encryption. The greater the sensitivity of the internal zones, the more restrictive the controls, to ensure that the 5 Ds of defense [2] can be applied:

1. **D**eter.
2. **D**etect.
3. **D**elay.
4. **D**isrupt.
5. **D**efend.

The very design of a Zero Trust model helps an organization to quickly find (**D**etect) the **ABNORMAL** from the NORMAL, so that quickly and efficiently respond to potential malicious or impactful activities—helping them to slow down (**D**elay) or stop (**D**isrupt) any lateral movements from an unauthorized incursion, in the protection of their sensitive assets (**D**efend). In addition, the defenses get progressively more robust as the value and sensitivity of the internal zones increase (**D**eter).

Consequently, the Zero Trust model is created from the following five assertions:

1. The architecture is assumed to be always under attack.
2. Both external and internal threat actors are persistently present.
3. Architecture environment is not sufficient for assuming trust in environment.
4. Every asset (e.g., user, device, traffic flow, etc.) is strictly authenticated and authorized.
5. Access policies are dynamic and calculated from as many data resources as possible.

Unlike the traditional security model (such as required by the payment card industry data security standard), as depicted in Fig. 1 [3], the network infrastructure employs a far more distributed policy enforcement to apply the Zero Trust principles, as depicted in Fig. 2.

As you can see, in the Zero Trust model, the supporting system is the Control plane, which coordinates and configures the data planes, ensuring that access requests for any asset is strictly authenticated and authorized. The greater the sensitivity, the greater the authentication and authorization, through dynamic configuration of the data planes to accept the traffic flows for specific clients.

Consequently, although Zero Trust models are more robust, they may not be suitable for every organization and some may choose to implement a hybrid model, where secure silos are created (using the Zero Trust model) inside a traditional security architecture model.

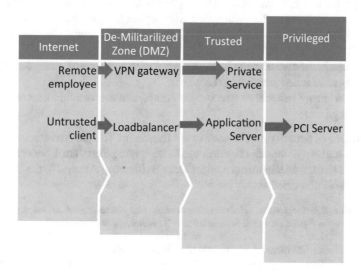

Fig. 1 Traditional network security architecture

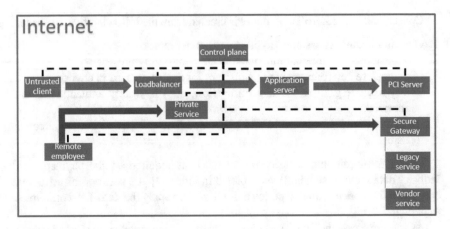

Fig. 2 Zero Trust Architecture

2 What Is Zero Trust?

There are a number of origins to the term 'Zero Trust', the first being associated with Stephen Marsh's 1984 University of Sterling Paper "Formalising Trust as a Computational Concept" [4] and the second being a reference to Ronald Reagan's approach to the Cold War [5].

Despite the concept being almost forty years old, it has now become a 'buzz term' which is not unusual for the security industry, as rather than reinvent the wheel it is easier to keep the same old wheel turning but just to add a new shiny new coat of paint.

Underneath, the construction is the same but with a new color, shiny coat of paint it helps to create a new interest in the wheel.

There are many references that supply comprehensive insights into what 'Zero Trust' is, e.g.,

- In Chap. 3 of Security and Privacy in the Internet of Things: Architectures, Techniques, and Applications (Egerton et al., "Applying Zero Trust Security Principles to Defence Mechanisms against Data Exfiltration Attacks"), the Zero Trust principal is described as:

    The explicit verification of the authentication and authorization of all actions is performed, regardless of the requesting user's credentials or permissions.

- The National Institute of Standards and Technology (NIST) Special Publication 800–207—Zero Trust Architecture [6] describes Zero Trust (ZT) as being:

    The term for an evolving set of cybersecurity paradigms that move defenses from static, network-based perimeters to focus on users, assets, and resources.

    A zero-trust architecture (ZTA) uses zero trust principles to plan industrial and enterprise infrastructure and workflows.

Zero trust assumes there is no implicit trust granted to assets or user accounts based solely on their physical or network location (i.e., local area networks versus the internet) or based on asset ownership (enterprise or personally owned).

Authentication and authorization (both subject and device) are discrete functions performed before a session to an enterprise resource is established.

Zero trust is a response to enterprise network trends that include remote users, bring your own device (BYOD), and cloud-based assets that are not located within an enterprise-owned network boundary.

Zero trust focuses on protecting resources (assets, services, workflows, network accounts, etc.), not network segments, as the network location is no longer seen as the prime component to the security posture of the resource.

- Gartner [7] summarize Zero Trust as beginning with just two projects:

Project 1: Zero trust network access (ZTNA)

In the past, when users left the "trusted" enterprise network, VPNs were used to extend the enterprise network to them. If attackers could steal a user's credentials, they could easily gain access to the enterprise network.

Zero trust network access abstracts and centralizes access mechanisms so that security engineers and staff can be responsible for them. It grants appropriate access based on the identity of the humans and their devices, plus other context such as time and date, geolocation, historical usage patterns and device posture. The result is a more secure and resilient environment, with improved flexibility and better monitoring.

Project 2: Identity-based segmentation

Identity-based segmentation, also known as micro or zero trust segmentation, is an effective way to limit the ability of attackers to move laterally in a network once they have gotten in.

Identity-based segmentation reduces excessive implicit trust by allowing organizations to shift individual workloads to a "default deny" rather than an "implicit allow" model. It uses dynamic rules that assess workload and application identity as part of determining whether to allow network communications.

The principle of the Zero Trust model is built around the belief that you should never trust and always verify and employs core principles, e.g., Microsoft [8] describe the following three principles built around technology pillars, as shown in Tables 1 and 2.

Whereas in the NIST SP800-207 [9], the core principles are described in Table 3.

TIKAJ [10] go onto to explain that the Zero Trust Model has seven focus areas are detailed in Table 4.

Table 1 Core principles

1. Verify explicitly	2. Use least privilege access	3. Assume breach
Always authenticate and authorize based on all available data points	Limit user access with Just-In-Time and Just-Enough-Access (JIT/JEA), risk-based adaptive policies, and data protection	Minimize blast radius and segment access. Verify end-to-end encryption and use analytics to get visibility, drive threat detection, and improve defenses

Table 2 Technology pillars

Identity	**Secure identity with Zero Trust**
	Identities—whether they represent people, services, or IoT devices—define the Zero Trust control plane. When an identity attempts to access a resource, verify that identity with strong authentication, and ensure access is compliant and typical for that identity. Follow least privilege access principles
Endpoints	**Secure endpoints with Zero Trust**
	Once an identity has been granted access to a resource, data can flow to a variety of different endpoints—from IoT devices to smartphones, BYOD to partner-managed devices, and on-premises workloads to cloud-hosted servers. This diversity creates a massive attack surface area. Monitor and enforce device health and compliance for secure access
Data	**Secure data with Zero Trust**
	Ultimately, security teams are protecting data. Where possible, data should remain safe even if it leaves the devices, apps, infrastructure, and networks the organization controls. Classify, label, and encrypt data, and restrict access based on those attributes
Applications	**Secure applications with Zero Trust**
	Applications and APIs provide the interface by which data is consumed. They may be legacy on-premises, lifted-and-shifted to cloud workloads, or modern SaaS applications. Apply controls and technologies to discover shadow IT, ensure appropriate in-app permissions, gate access based on real-time analytics, monitor for abnormal behavior, control user actions, and validate secure configuration options
Infrastructure	**Secure infrastructure with Zero Trust**
	Infrastructure—whether on-premises servers, cloud-based VMs, containers, or micro-services—represents a critical threat vector. Assess for version, configuration, and JIT access to harden defense. Use telemetry to detect attacks and anomalies, and automatically block and flag risky behavior and take protective actions
Network	**Secure networks with Zero Trust**
	All data is ultimately accessed over network infrastructure. Networking controls can provide critical controls to enhance visibility and help prevent attackers from moving laterally across the network. Segment networks (and do deeper in-network micro-segmentation) and deploy real-time threat protection, end-to-end encryption, monitoring, and analytics

The utilization of the Zero Trust model has been included in the United States President's Executive Order on Improving the Nation' Cybersecurity [11]:

Sec. 3. Modernizing Federal Government Cybersecurity.

a) *To keep pace with today's dynamic and increasingly sophisticated cyber threat environ-ment, the Federal Government must take decisive steps to modernize its approach to cybersecurity, including by increasing the Federal Government's visibility into threats, while protecting privacy and civil liberties. The Federal Government must adopt security best practices; advance toward Zero Trust Architecture; accelerate movement to secure cloud services, including Software as a Service (SaaS), Infrastructure as a Service (IaaS),*

Table 3 NIST SP800-207 Zero Trust Principles

1. Continuous verification	Always verify access, all the time, for all resources	*Continuous verification means no trusted zones, credentials, or devices at any time. Hence the common expression "Never Trust, Always Verify." Verification that must be applied to such a broad set of assets continuously means that several key elements must be in place for this to work effectively:* • *Risk based conditional access. This ensures the workflow is only interrupted when risk levels change, allowing continual verification, without sacrificing user experience* • *Rapid and scalable dynamic policy model deployment. Since workloads, data, and users can move often, the policy must not only account for risk, but also include compliance and IT requirements for policy. Zero Trust does not alleviate organizations from compliance and organizational specific requirements*
2. Limit the "blast radius"	Minimize impact if an external or insider breach occurs	*If a breach does occur, minimizing the impact of the breach is critical. Zero Trust limits the scope of credentials or access paths for an attacker, giving time for systems and people to respond and mitigate the attack Limiting the radius means:* • *Using identity-based segmentation. Traditional network-based segmentation can be challenging to maintain operationally as workloads, users, data, and credentials change often* • *Least privilege principle* – *Whenever credentials are used, including for non-human accounts (such as service accounts), it is critical these credentials are given access to the minimum capability required to perform the task. As tasks change, so should the scope. Many attacks leverage over privileged service accounts, as they are typically not monitored and are often overly permissioned*

3. Automate context collection and response	Incorporate behavioral data and get context from the entire IT stack (identity, endpoint, workload, etc..) for the most accurate	*To make the most effective and accurate decisions, more data helps so long as it can be processed and acted on in real-time. NIST provides guidance on using information from the following sources:*
		• *User credentials—human and non-human (service accounts, non-privileged accounts, privileged accounts—including SSO credentials)*
		• *Workloads—including VMs, containers, and ones deployed in hybrid deployments*
		• *Endpoint—any device being used to access data*
		• *Network*
		• *Data*
		• *Other sources (typically via APIs):*
		  – *SIEM*
		  – *SSO*
		  – *Identity providers (like AD)*
		  – *Threat intelligence*

Table 4 TIKAJ 7 focus areas for Zero Trust

• *Zero Trust networks*	*Attackers need to enter within your network to steal sensitive data. With Zero Trust networks it becomes extremely difficult, zero trust networks protect, segment, isolate and restrict the threats using the most revolutionary firewalls*
• *Zero Trust people*	*As we all know, humans can make mistakes, and hence they are the cause of most data breaches, mistakes can be anything from bad passwords, infrequent wrong clicks and can let anyone steal your credentials and impersonate you behind the screen. So, monitor your users' activity within your network and protect them*
• *Zero Trust devices*	*With everyone working remotely the number of devices within your network is increasing multifold and hence the need to secure the network is more important. Every connected device in your network is an entry point for the attackers and that can cause infiltration in your network. Your security team should be able to protect, isolate and have control over every connected device*
• *Zero Trust data*	*Data is the ultimate target of the attackers and even the inside threats as with data anyone can do anything. Hence, it becomes more important for the security team to protect data before it becomes a breach. Your organization should have the option to know where the information resides, who can access it, what's sensitive, and screen information to distinguish and react to possible dangers*
• *Zero Trust workloads*	*The workload is running on a public cloud, which makes it even more difficult to detect and defend against dangers as it is a dynamic infrastructure and backend software that enables communication between the network and users*
• *Zero Trust visibility and analytics*	*This principle suggests the use of advanced technologies such as AI to automate the detection, protection, encryption processes which might anomaly detection and configuration control. Advanced threat detection, user behavior analytics and identify potential threats so that you can identify anomalous behavior*
• *Zero Trust automation and orchestration*	*With such a fast-paced life, no one is having time to wait for the threat to happen and then rectify it. With everything across being so dynamic it becomes even more difficult to detect threats and to make your network secure, you need to examine quickly and automate the whole zero trust system centrally*

and Platform as a Service (PaaS); centralize and streamline access to cybersecurity data to drive analytics for identifying and managing cybersecurity risks; and invest in both technology and personnel to match these modernization goals.

b) *Within 60 days of the date of this order, the head of each agency shall:*

   i. *update existing agency plans to prioritize resources for the adoption and use of cloud technology as outlined in relevant OMB guidance;*

   ii. *develop a plan to implement Zero Trust Architecture, which shall incorporate, as appropriate, the migration steps that the National Institute of Standards and Technology (NIST) within the Department of Commerce has outlined in standards and guidance, describe any such steps that have already been completed, identify activities that will have the most immediate security impact, and include a schedule to implement them; and*

   iii. *provide a report to the Director of OMB and the Assistant to the President and National Security Advisor (APNSA) discussing the plans required pursuant to subsection (b)(i) and (ii) of this section.*

c) *As agencies continue to use cloud technology, they shall do so in a coordinated, deliberate way that allows the Federal Government to prevent, detect, assess, and remediate cyber incidents. To facilitate this approach, the migration to cloud technology shall adopt Zero Trust Architecture, as practicable. The CISA shall modernize its current cybersecurity programs, services, and capabilities to be fully functional with cloud-computing environments with Zero Trust Architecture. The Secretary of Homeland Security acting through the Director of CISA, in consultation with the Administrator of General Services acting through the Federal Risk and Authorization Management Program (FedRAMP) within the General Services Administration, shall develop security principles governing Cloud Service Providers (CSPs) for incorporation into agency modernization efforts. To facilitate this work:*

   i. *Within 90 days of the date of this order, the Director of OMB, in consultation with the Secretary of Homeland Security acting through the Director of CISA, and the Administrator of General Services acting through FedRAMP, shall develop a Federal cloud-security strategy and provide guidance to agencies accordingly. Such guidance shall seek to ensure that risks to the FCEB from using cloud-based services are broadly understood and effectively addressed, and that FCEB Agencies move closer to Zero Trust Architecture.*

   ii. *Within 90 days of the date of this order, the Secretary of Homeland Security acting through the Director of CISA, in consultation with the Director of OMB and the Administrator of General Services acting through FedRAMP, shall develop and issue, for the FCEB, cloud-security technical reference architecture documentation that illustrates recommended approaches to cloud migration and data protection for agency data collection and reporting.*

   iii. *Within 60 days of the date of this order, the Secretary of Homeland Security acting through the Director of CISA shall develop and issue, for FCEB Agencies, a cloud-service governance framework. That framework shall identify a range of services and protections available to agencies based on incident severity. That framework shall also identify data and processing activities associated with those services and protections.*

   iv. *Within 90 days of the date of this order, the heads of FCEB Agencies, in consultation with the Secretary of Homeland Security acting through the Director of CISA, shall evaluate the types and sensitivity of their respective agency's unclassified data, and shall provide to the Secretary of Homeland Security through the Director of CISA and to the Director of OMB a report based on such evaluation. The evaluation shall prioritize identification of the unclassified data considered by the agency to be the most sensitive and under the greatest threat, and appropriate processing and storage solutions for those data.*

d) *Within 180 days of the date of this order, agencies shall adopt multi-factor authentication and encryption for data at rest and in transit, to the maximum extent consistent with Federal records laws and other applicable laws. To that end:*

  i. *Heads of FCEB Agencies shall provide reports to the Secretary of Homeland Security through the Director of CISA, the Director of OMB, and the APNSA on their respective agency's progress in adopting multifactor authentication and encryption of data at rest and in transit. Such agencies shall provide such reports every 60 days after the date of this order until the agency has fully adopted, agency-wide, multi-factor authentication and data encryption.*

  ii. *Based on identified gaps in agency implementation, CISA shall take all appropriate steps to maximize adoption by FCEB Agencies of technologies and processes to implement multifactor authentication and encryption for data at rest and in transit.*

  iii. *Heads of FCEB Agencies that are unable to fully adopt multi-factor authentication and data encryption within 180 days of the date of this order shall, at the end of the 180-day period, provide a written rationale to the Secretary of Homeland Security through the Director of CISA, the Director of OMB, and the APNSA.*

e) *Within 90 days of the date of this order, the Secretary of Homeland Security acting through the Director of CISA, in consultation with the Attorney General, the Director of the FBI, and the Administrator of General Services acting through the Director of FedRAMP, shall establish a framework to collaborate on cybersecurity and incident response activities related to FCEB cloud technology, in order to ensure effective information sharing among agencies and between agencies and CSPs.*

f) *Within 60 days of the date of this order, the Administrator of General Services, in consultation with the Director of OMB and the heads of other agencies as the Administrator of General Services deems appropriate, shall begin modernizing FedRAMP by:*

  i. *establishing a training program to ensure agencies are effectively trained and equipped to manage FedRAMP requests, and providing access to training materials, including videos-on-demand;*

  ii. *improving communication with CSPs through automation and standardization of messages at each stage of authorization. These communications may include status updates, requirements to complete a vendor's current stage, next steps, and points of contact for questions;*

  iii. *incorporating automation throughout the lifecycle of FedRAMP, including assessment, authorization, continuous monitoring, and compliance;*

  iv. *digitizing and streamlining documentation that vendors are required to complete, including through online accessibility and pre-populated forms; and*

  v. *identifying relevant compliance frameworks, mapping those frameworks onto requirements in the FedRAMP authorization process, and allowing those frameworks to be used as a substitute for the relevant portion of the authorization process, as appropriate.*

However, the concept of Zero Trust it is important to note that this should not be only regarded as a cybersecurity/network security model and that it should be noted that this can also be applied to the enhancement of an organization's physical security measures.

3   Enhancing Physical Security with Zero Trust

Many corporate businesses employ electronic automated access control systems (EAACS) to restrict access at the perimeter and limited access restrictions for internal trusted zones—traditional security. This type of model relies on the trust that everyone inside the corporate physical environment are trustworthy individuals and that the perimeter controls are robust enough to prevent a threat actor from gaining unauthorized lateral movement between internal trusted environments. Much the same as the traditional cyber security model, this will rely on a defense in depth [12] approach, where concentric layers of defense should help to protect an organization's sensitive or valued assets from unauthorized access or compromise, as depicted in Fig. 3.

However, the reality is that should the outer perimeter be compromised, it is often the case that the inner defensive layers can be easily peeled back (compromised) to allow an attacker to gain access to the sensitive or valued assets, deeper inside the core of an establishment.

With the Zero Trust approach [13], each layer requires robust access verification, the application of the principle of least privilege [14] and that each layer could be compromised.

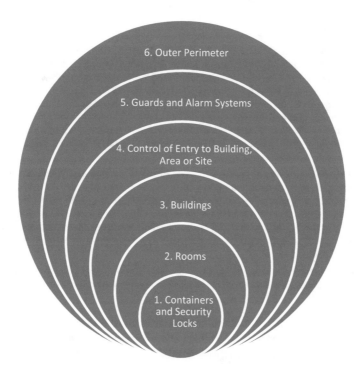

Fig. 3   Defense in Depth (DiD)

The use of the Zero Trust approach becomes extremely relevant and important when and organization perceives there to be ever present threat to their critical assets and that there would be a significant impact should these critical assets become compromised by one or more of these threat actors. Consequently, during the Cold War conflict (1947–1991) [15], when after the conclusion of World War Two there was a forty-year conflict between the Eastern and Western Global Superpowers.

4   Network Architecture Design

Despite the European Union General Data Protection Regulation (EU GDPR) introducing the requirements for 'Secure by Design' and 'Secure by Default' operations (Article 25), many businesses do not successfully adopt and implement such practices.

4.1   Art. 25 GDPR: Data Protection by Design and by Default [16]

1. *Taking into account the state of the art, the cost of implementation and the nature, scope, context and purposes of processing as well as the risks of varying likelihood and severity for rights and freedoms of natural persons posed by the processing, the controller shall, both at the time of the determination of the means for processing and at the time of the processing itself, implement appropriate technical and organisational measures, such as pseudonymisation, which are designed to implement data-protection principles, such as data minimisation, in an effective manner and to integrate the necessary safeguards into the processing in order to meet the requirements of this Regulation and protect the rights of data subjects.*

2. *The controller shall implement appropriate technical and organizational measures for ensuring that, by default, only personal data which are necessary for each specific purpose of the processing are processed. That obligation applies to the amount of personal data collected, the extent of their processing, the period of their storage and their accessibility. In particular, such measures shall ensure that by default personal data are not made accessible without the individual's intervention to an indefinite number of natural persons.*

3. *An approved certification mechanism pursuant to Article 42 may be used as an element to demonstrate compliance with the requirements set out in paragraphs 1 and 2 of this Article.*

Where network-based EU data subject operations are being conducted, the implementation and maintenance of a suitable network architecture should be considered as a fundamental contribution to this article.

'Zero Trust Architecture (ZTA)' is only one of the network architectural models/frameworks that are available for adoption. However, some of the more traditional network architectures (when implemented correctly might be more proper for your organization and, so, other security models [17, pp. 295–346]) should not be discounted, e.g.,

- **Bell-LaPadula.**
- **Biba.**
- **Clark–Wilson.**
- **Lattice-Based.**
- **Cisco SAFE.**

5 Zero Trust in the Cold War Era

From the post war, several military bases have been designed and developed to defend against the ever-present Cold War [18] threats and the greater the threats, the greater the need for a Zero Trust approach. No longer would a reliance on a robust perimeter be considered as being sufficient to counter the threats. Consequently, the security measures were designed and developed using the pretense that the enemy would be able to breach the perimeter and within the various defensive layers' verification, and damage limitation, measures would be needed.

I started my career in Royal Air Force, in the late nineteen eighties, when the Cold War threat was still seen as being a real threat to the military and nuclear weapons were still regarded as a suitable defense. As a result, as a Royal Air Force Policeman, my basic training would include (in addition to the seven-week basic police training) six-weeks of Special Duties 814 training so that I could be deployed to help protect the numerous high-importance air bases [19], weapons storage facilities [20] and nuclear bunkers [21].

In such facilities, the bases were designed [22, 23] using the principle of 'Zero Trust' (*although it was never called this, at the time*). Later, the importance of 'Zero Trust' became even more important for the design and development of temporary, deployed military bases [24].

Despite the fact that individuals may have been issued with an official military identity card (e.g., MOD 90), this did not give them the automatic rights to gain entry into anywhere, on any military base and even an individual's seniority in rank did not automatically give them the rights for the 'Need to Know'/'Need to Access' to a restricted or sensitive area. In fact, there was a well-known phrase the was often used by many a member of the RAF Police:

With all due respect, Sir/Ma'am, please do not confuse your rank with my authority.

Even being a member of the RAF Police did not allow the automatic rights to the 'Need to Know'/'Need to Access' to sensitive areas. Such rights needed to be explicitly approved and were timebound, following legitimate needs, e.g.,

- **Just-In-Time** (JIT) [25].

  JIT access helps organizations provision access so that users only have the privileges to access privileged accounts and resources when they need it, and not otherwise any other times.

  Instead of granting always-on (or standing) access (or standing access), organizations can use JIT access to limit access to a specific resource for a specific timeframe.

  This granular approach mitigates the risk of privileged account abuse by significantly reducing the amount of time a cyber attacker or malicious insider has to gain access to privileged accounts before moving laterally through a system and gaining unauthorized access to sensitive data.

- **Just-Enough-Access (JEA)** [26].

  Ensures that only those privileges needed to carry out the request are provided for the duration of the request.

Added measures would be needed, the closer someone got to the highly restricted/highly sensitive areas or assets. For example, where a highly restricted aircraft was parked up on an aircraft pan (as depicted in Fig. 4 [27]), this area would have a resident guard checking that any visiting personnel were employing the two-person principle and that both of the visitors were listed on the timebound access control list (*if their names weren't on the list or their identities could not be verified, access would not be granted*) and where an RAF establishment had a special weapons storage facility (as depicted in Fig. 5 [28, 29]) or hardened aircraft shelters (as depicted in Fig. 6 [30, 31]) within its confines, having access did not automatically give access to the special weapons storage facility or the hardened aircraft shelter dispersals.

Every person accessing these areas would go through rigorous verification checks, whilst the facilities would have robust defenses, supported by regular exercises that would prepare the security teams for breach or compromise situations.

These became hardened security bases ('Micro-Bases'), inside established RAF bases, having their own secured ingress and egress routes through the outer defensive layers of the parent RAF base.

Even if the RAF base did not have permanent 'Micro-Bases', in defense of the Cold War threats, every RAF base would have the capability to quickly ramp up their defenses to adopt a Zero Trust approach. The effectiveness of their Zero Trust approaches would be subject to independent validation, through regular RAF station exercises (e.g., TACEVAL [32], MAXEVAL [33], MINEVAL [34], etc.).

Additionally, such establishments would still have permanent areas where the 'Zero Trust's' guiding principles [35] would be applied, e.g.,

Fig. 4 Cold War aircraft

- **Verify access and Least Privilege**
  - Six-hourly out of hours checks.

    Checking security container check sheets, secure offices, buildings, etc.

  - 24/7 centralized command post.
  - Access controls.
  - Identity checks.
  - Monitor EAACS.

- **Assume breach**
  - Checking integrity of security containers, secure offices, buildings, etc.
  - Monitor alarms and closed-circuit television (CCTV).
  - Periodic security patrols.

    Internal.
    External.

  - Incident response.
  - Security integrity checks.
  - Security reports.

Fig. 5 Special weapons storage facility

Fig. 6 Hardened aircraft shelter dispersal

6 Implementing a Zero Trust Approach

Although I believe that there are very few businesses that are suitable for a fully 100% Zero Trust model, there are elements of this that could be beneficial enhancements to your organization's cyber (network) and physical security models [36].

In essence, a true 'Zero Trust' model completely eradicates the need for convenience. Before considering whether a 'Zero Trust' model is right for you, you need to understand your existing infrastructures (both network and physical), by answering the **6 WH** questions:

- What are your restrictions/limitations to the adoption of a 'Zero Trust' model?
- Which assets/operations are important to your organization?
- Where do these assets/operations live?
- Who needs access to these assets/operations?
- Why do the need access?
- When do the need access?
- How will the access be explicitly authorized?
- How can these high-value/sensitive assets/operations be isolated from the less valuable/less-sensitive assets/operations?
- How are the assets linked/connected/associated?

Only when you understand and appreciate the **6WHs** will you be in a suitable place to begin to understand and appreciate which assets are involved with supporting these high value/sensitive business operations/processes. Consider the benefits of using automation (such as network mapping) to help identify these assets and to record them in an asset inventory.

Having answered the **6WH** questions, next you will need to visualize the supporting environments using network and physical topology and data/process flow diagrams.

Your topology diagrams and asset inventory are essential to help you to show the segmentation points, which assets are allowed to interact with each other (*Explicitly verified*), and which should be prohibited (*Least Privilege*) and where the points of ingress/egress (both physical and network) are found. This is where convenience should be eradicated so that traffic flows are strictly controlled and the use of privileged user accounts strictly controlled and only used when they are absolutely needed, to achieve legitimate business requirements.

Once you have set up the foundations and ground rules needed for 'Zero Trust', next, you need to dedicate considerable effort into monitoring the situation, ensuring that people stick to these requirements and that the supporting infrastructure continues to enforce the 'Zero Trust' principles. Where possible, consider the benefits of employing automation orchestration so that strictly defined policies and artificial intelligence (AI) or machine learning (ML) help to quickly show any non-conformities to the 'Zero Trust' Principles.

Let's face it, people are creature of habit and will quickly fall back into what they have been used to, or will seek the easiest or most convenient path for them (e.g., Libidinal Economy [37]) so you should be actively monitoring for the presence of the **ABNORMAL** from the expected NORMAL activities that a 'Zero Trust' model would bring.

Finally, assuming that you will suffer a non-conformity (Breach) of the 'Zero Trust' principles it is important to ensure that you have effective and well trained incident response operations so that you are able to quickly identify and respond to any non-conformities before they are able to impact or do damage to your valued business operations.

Underlying all of these, is the need for governance (setting the 'Tone at The Top'), policies, standard operating procedures, communication, training (onboarding and periodic refresher), audit and a suitable disciplinary process.

7   Conclusion

There is plenty of 'buzz' around the term 'Zero Trust', as an approach that businesses can adopt to reduce their risks of becoming the next victim of a cyber-attack. Even global governments (such as the United States) are endorsing and encouraging organizations to adopt and implement the 'Zero Trust' model.

However, I would argue that for many businesses this model is a 'step too far' and they should be encouraged to start by naming a suitable security model, which they are capable of correctly implementing and supporting. Before jumping to the conclusion that a breached organization's network architecture security model was incorrect, it would be helpful to understand whether their chosen security model was running as intended. For example, what would be the response to questions, such as:

- Do you understand where your 'Crown Jewels' resided?
- What assets are needed to interact with the 'Crown Jewels' estate?
- Had you conducted business impact analysis (BIA) exercises of your business operations, to categorize and prioritize your assets?
- What security model were you using?
- Did you apply the 'Least Privilege' principle?
- Did you regularly evaluate the effectiveness of your security model?
- What were your identified risks, associated with the category of business operation?

Despite this now being a legal requirement (about the personal data operations, for European Union and United Kingdom data subjects), rarely has a breached business implemented a 'Secure by Design', supported a 'Secure by Default' supporting infrastructure.

Instead, they will often be reliant on the infrastructure perimeter defenses to provide all their protection from the 'Badlands' (Internet) and, as a consequence, it only takes a single mis-configured device, residing at the perimeter to allow a malicious individual to gain unauthorized access to whatever lies beyond (*often referred to as being a flat network* [38]), by having the opportunity to gain a clandestine presence inside a target's corporate infrastructure so that they can freely (*and without detection*) move laterally across their victims infrastructure.

Additionally, such an approach supplies little protection against the insider threat (*be that a deliberate or accidental action*).

Before deciding to jump straight into designing and implementing the 'Zero Trust' model, it is important to ensure that this is compatible with your business model and a better approach might be to start by getting the basics right before moving to a more mature 'Zero Trust' infrastructure [39].

References

1. Cybersecurity & Infrastructure Security Agency (CISA). Shields up | CISA. www.cisa.gov, www.cisa.gov/shields-up
2. Shertz K (2018) The five Ds: securing the nation's aviation infrastructure. 23 Mar 2018
3. Gilman E, Barth D (2017) Zero trust networks: building secure systems in untrusted networks. O'reilly Media, Sebastopol, CA
4. Marsh S (1994) Formalising trust as a computational concept
5. Poe M (2021) Building a zero trust security plan for your company—BeyondID, Inc. BeyondID, 29 Mar 2021. www.beyondid.com/resources/zero-trust-security-plan. Accessed 25 Feb 2022
6. Rose S et al (2020) Zero trust architecture, 11 Aug 2020. https://doi.org/10.6028/nist.sp.800-207. nvlpubs.nist.gov/nistpubs/SpecialPublications/NIST.SP.800-207.pdf
7. Moore S (2021) How to get started with zero trust security. Gartner, 3 June 2021. www.gartner.com/smarterwithgartner/new-to-zero-trust-security-start-here
8. Centric G et al (2022) Zero trust guidance center. Docs.microsoft.com, 8 Feb 2022. docs.microsoft.com/en-us/security/zero-trust/#guiding-principles-of-zero-trust. Accessed 24 Feb 2022
9. Raina K (2022) What is zero trust security? Principles of the zero trust model. Crowdstrike.com, 6 May 2021. www.crowdstrike.com/cybersecurity-101/zero-trust-security. Accessed 24 Feb 2022
10. Srivastava A (2021) 7 principles of zero trust security: break the implicit trust biases! | TIKAJ. www.tikaj.com, 18 Oct 2021. www.tikaj.com/blog/zero-trust-security-break-the-implicit-trust. Accessed 25 Feb 2022
11. United States White House. Executive order on improving the nation's cybersecurity. The White House, 12 May 2021. www.whitehouse.gov/briefing-room/presidential-actions/2021/05/12/executive-order-on-improving-the-nations-cybersecurity. Accessed 24 Feb 2022
12. robmazz. Datacenter physical access security—Microsoft Service Assurance. Docs.microsoft.com, 17 Nov 2021. docs.microsoft.com/en-us/compliance/assurance/assurance-datacenter-physical-access-security. Accessed 24 Feb 2022
13. Egerton H et al (2021) Applying zero trust security principles to defence mechanisms against data exfiltration attacks. Security and privacy in the Internet of Things, 3 Dec 2021, pp 57–89. https://doi.org/10.1002/9781119607755.ch3. Accessed 24 Feb 2022
14. CSRC NIST Glossary. Least privilege—glossary | CSRC. Csrc.nist.gov, NIST. csrc.nist.gov/glossary/term/least_privilege
15. The Editors of Encyclopedia Britannica (2020) Cold War | Causes, facts, & summary | Britannica. Encyclopedia Britannica. www.britannica.com/event/Cold-War#ref284221
16. GDPR.EU (2018) Art. 25 GDPR—data protection by design and by default. GDPR.eu, 14 Nov. 2018. gdpr.eu/article-25-data-protection-by-design. Accessed 28 Feb 2022

17. Seaman J (2021) Protective security: creating military-grade defenses for your digital business. Apress, S.L., pp 295–346
18. Schofield J et al (2021) Cold War: a transnational approach to a global heritage. Post-Mediev Archaeol 55(1):39–58. https://doi.org/10.1080/00794236.2021.1896211. Accessed 29 Dec 2021
19. The Guardian (2014) Revealed: the 106 Cold War nuclear targets across the UK. Rob Edwards, 5 June 2014. www.robedwards.com/2014/06/revealed-the-106-cold-war-nuclear-targets-across-the-uk.html. Accessed 25 Feb 2022
20. Tolley S. RAF Barnham nuclear weapon storage site. rafbarnham-nss.weebly.com. Accessed 25 Feb 2022
21. Subterranea Brittanica. Nuclear bunkers—Subterranea Britannica. www.subbrit.org.uk, www.subbrit.org.uk/categories/nuclear-bunkers. Accessed 25 Feb 2022
22. Department of Defense (DoD) (1993) Military handbook design guidelines for physical security of facilities
23. Headquarters Department of the Army (2017) Army Publishing Directorate. Army-pubs.army.mil, Jan 2017. armypubs.army.mil/ProductMaps/PubForm/Details.aspx?PUB_ID= 1001631. Accessed 26 Feb 2022
24. U.S. Army Engineer Research and Development Center (2004) Base camp protection and survivability demonstration program
25. CyberArk. Just-in-time access. CyberArk. www.cyberark.com/what-is/just-in-time-access
26. Shore M et al (2021) Zero trust: the what, how, why, and when. Computer 54(11):26–35. https://doi.org/10.1109/mc.2021.3090018. Accessed 28 Oct 2021
27. Imperial War Museums. The Royal Air Force, 1950–1969. Imperial War Museums. www.iwm.org.uk/collections/item/object/205214210. Accessed 26 Feb 2022
28. Google Earth. Earth.google.com, 22 Apr 2021. earth.google.com/web/search/RAF+Honing ton/@52.34861308. Accessed 26 Feb 2022
29. Jackson B (2008) RAF Faldingworth. Flickr, 27 Apr 2008. www.flickr.com/photos/sd814/with/2420494143. Accessed 26 Feb 2022
30. Google Earth (2021) Google Earth. Earth.google.com, 22 Apr 2021. earth.google.com/web/search/RAF+Honington/@52.3479183. Accessed 26 Feb 2022
31. Ramsden G (2022) Panavia Tornado GR1T, ZA357/BT010, Royal Air Force: Abpic.co.uk, 23 Sept 2017. abpic.co.uk/pictures/view/1578695. Accessed 26 Feb 2022
32. Schroeder W (2019) NATO at seventy: filling NATO's critical defense-capability gaps. Atlantic Council, 4 Apr 2019. www.atlanticcouncil.org/in-depth-research-reports/report/nato-at-sev enty-filling-nato-s-critical-defense-capability-gaps. Accessed 26 Feb 2022
33. Air R, Air R (1999) Royal Air Force in Germany, 1945–1993. Royal Air Force Historical Society, Newcastle
34. Franklin M (2012) Mission accomplished: my RAF stations, squadrons and air-craft. Mission Accomplished, 10 June 2012. missionaccom-plished23.blogspot.com/2012/06/my-raf-stations-squadrons-and-aircraft.html. Accessed 26 Feb 2022
35. Adams S, Banasik TJ. The 6 pillars of zero trust
36. Jain A (2022) Why the zero trust model matters now for physical security. Vector Flow, 5 Aug 2021. vectorflow.com/blog/why-the-zero-trust-model-matters-now-for-cyber-and-phy sical-security. Accessed 27 Feb 2022
37. Bennett D (2013) Currency of desire: libidinal economy, psychoanalysis and sexual revolution. Lawrence & Wishart Ltd
38. Rubens P (2021) The risks and rewards of flat networks. Enterprise Networking Planet, 27 Apr 2021. www.enterprisenetworkingplanet.com/data-center/the-risks-and-rewards-of-flat-networks. Accessed 28 Feb 2022
39. Ghosh D (2012) NIST's guidance for a zero trust architecture: roadmap for deploying an enterprise security model zero trust architecture. ManageEngine